CAN THE SEAS SURVIVE US?

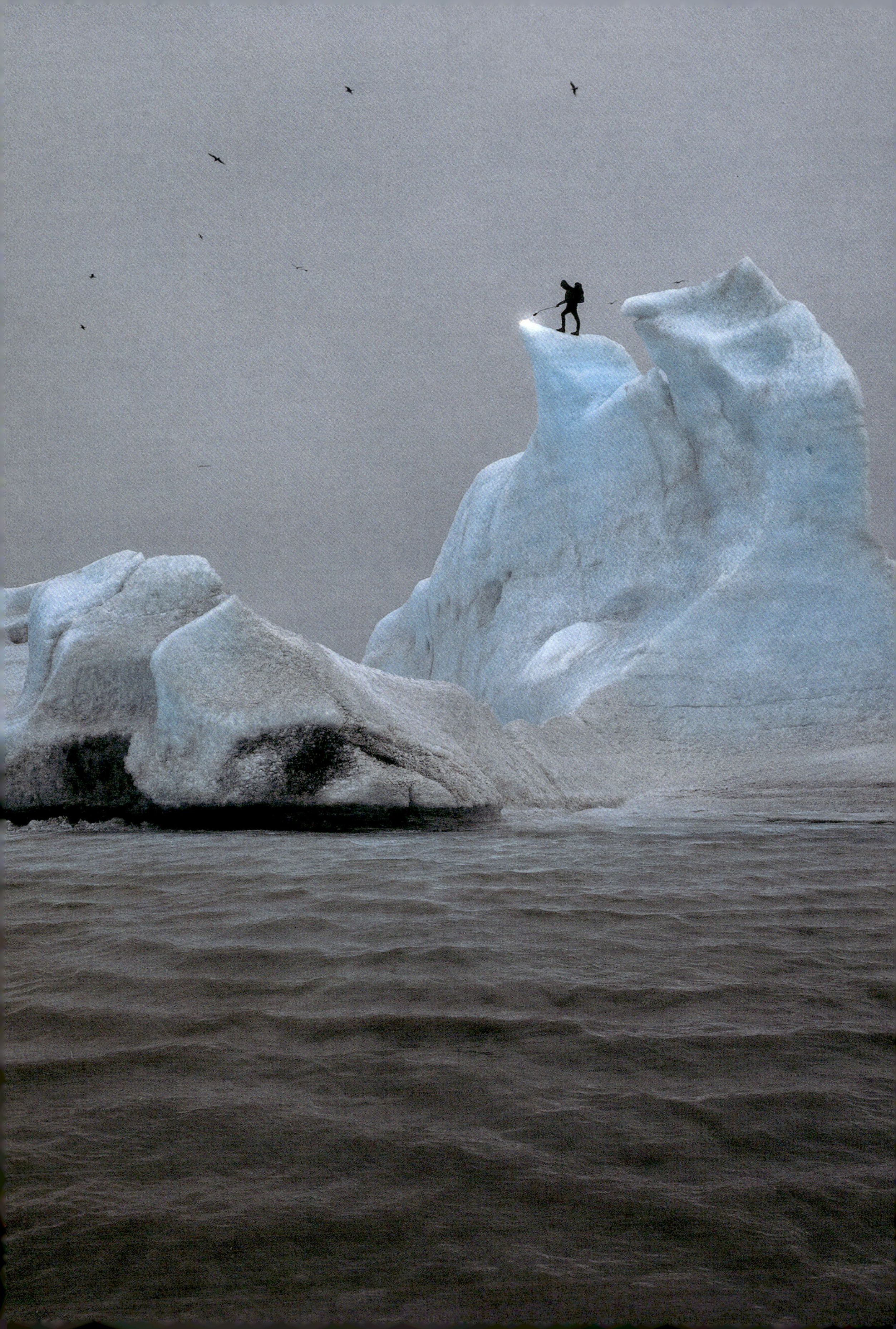

CAN THE SEAS SURVIVE US?

EDITED BY TANIA MOORE AND
JOHN KENNETH PARANADA

Published to accompany the *Can the Seas Survive Us?* season of exhibitions at the Sainsbury Centre, University of East Anglia, Norwich, Norfolk:

A World of Water, 15 March–3 August 2025
Darwin in Paradise Camp: Yuki Kihara, 15 March–3 August 2025
Sea Inside, 7 June–21 September 2025

https://www.sainsburycentre.ac.uk

First published in 2025 by Kulturalis Ltd
14 Old Queen Street, London SW1H 9HP
United Kingdom
www.kulturalis.com

ISBN 978-1-83636-005-6

A catalogue record for this book is available from the British Library

Series Editor: Tania Moore
Project Manager: Lizzy Silverton
Design: James Alexander, Jade Design
Printed and bound in Turkey
Reproduction by Opero, Verona

10 9 8 7 6 5 4 3 2 1

Front cover: Julian Charrière, *Midnight Zone*, 2024
p. 2: Julian Charrière, *The Blue Fossil Entropic Stories III*, 2013

CONTENTS

FOREWORD
JAGO COOPER

Has nature been overtaken by culture; the natural world overwhelmed by a cultural one? Looking through the eyes of the extraordinary artists within this book provides new imaginations with which to answer that question. Not only do they allow us to reconceptualise the worlds of water that surround us, but they also take us with them, to venture beneath the surface.

The oceans cover more than 70 per cent of the planet; the epitome of the natural world, an eternal expanse of more than a billion cubic kilometres of marine habitat. Yet what do these seas say about the state of nature today and our relationship with it? From wind farms populating the North Sea and plastics filling the Pacific, to algal blooms appearing and beautiful creatures disappearing, what really lies below the surface of these stories and ultimately: *Can the Seas Survive Us?*

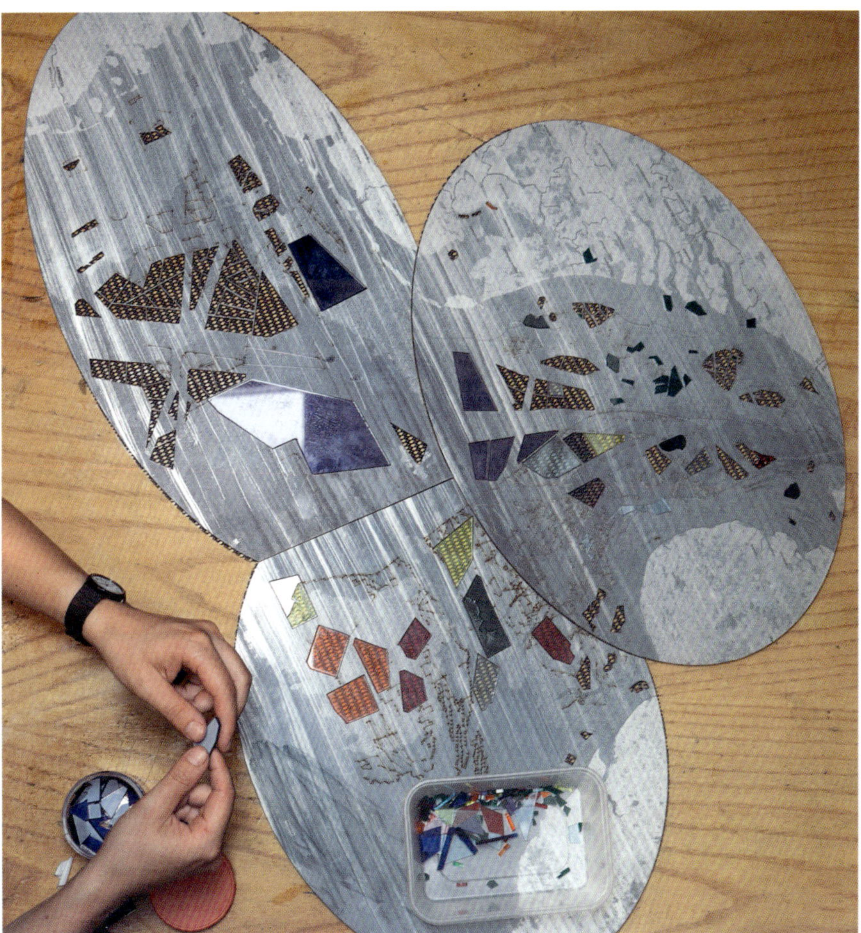

Fig. 0.0
Arieh Frosh and Ed Compson, *A Guide to Offshore Wind Farm Mapping*, 2024. Norwich University of the Arts

In planning for this wider artistic project, several members of staff from the Sainsbury Centre sailed across the North Sea, together with a collective of artists, academics and activists. We were journeying to meet friends and colleagues in the Netherlands who have been deliberating the same questions that we have, but often approaching them from a different angle. As we travelled on a 100-year-old East Anglian tall ship, the daytime view of endless grey waves transformed into a dazzling night-time horizon of coloured lights as the wind farms, cargo ships, oil rigs, floating platforms and buoys surrounded our tiny old wooden boat on all sides. The perception of the sea by those onboard swayed back and forth throughout the voyage between natural wonder and technological awe, a tension explored beautifully by the artworks discussed in the pages that follow.

In gathering artists, not only from the UK and the Netherlands, but from all around the world, the diversity of the lens through which the issues of the sea are perceived is incredibly revealing. Whales calling from bathtubs and puzzles of wind-farm pieces appear across timelines that collapse a historical world of Nelson with alternative visions of our possible futures. The trajectory of our relationship with the planet is illuminated here for all to find their own view.

INTRODUCTION: CAN THE SEAS SURVIVE US?

JOHN KENNETH PARANADA

Over the past five centuries human activity has dramatically transformed the Earth's landscape and disrupted its delicate water systems. The burning of fossil fuels and the expansion of extractive colonial capitalism have resulted in irreversible environmental damage and social upheaval. From the rise of empires and exploitation through slavery to the destruction of ancestral lands and the displacement of Indigenous communities, the relentless pull on our natural resources has led to what we now recognise as the Anthropocene – a time when human influence has permanently reshaped the planet's ecosystems.

As we face the urgent question, *Can the Seas Survive Us?*, it is essential to reflect on key global initiatives, including the Kyoto Protocol,[1] the Paris Agreement[2] and the United Nations Sustainable Development Goals (SDGs) for 2030.[3] With less than a decade left to meet these objectives – particularly SDG 14, which focuses on protecting life below water[4] – the call for collective action is critical.

Recent findings from the UN's Intergovernmental Panel on Climate Change (IPCC), alongside the momentum of COP26 and COP27, have highlighted the fundamental role of cultural institutions, museums and heritage sites in encouraging reflection and inspiring a shift towards sustainability.[5] Climate justice, human rights, political equality and recognising the intrinsic rights of nature are not just aspirational goals but pressing necessities if we are to deliver on the SDG 2030 agenda and confront the growing climate emergency.

Fig. 0.1
Julian Charrière, *Pure Waste*, 2021, film still

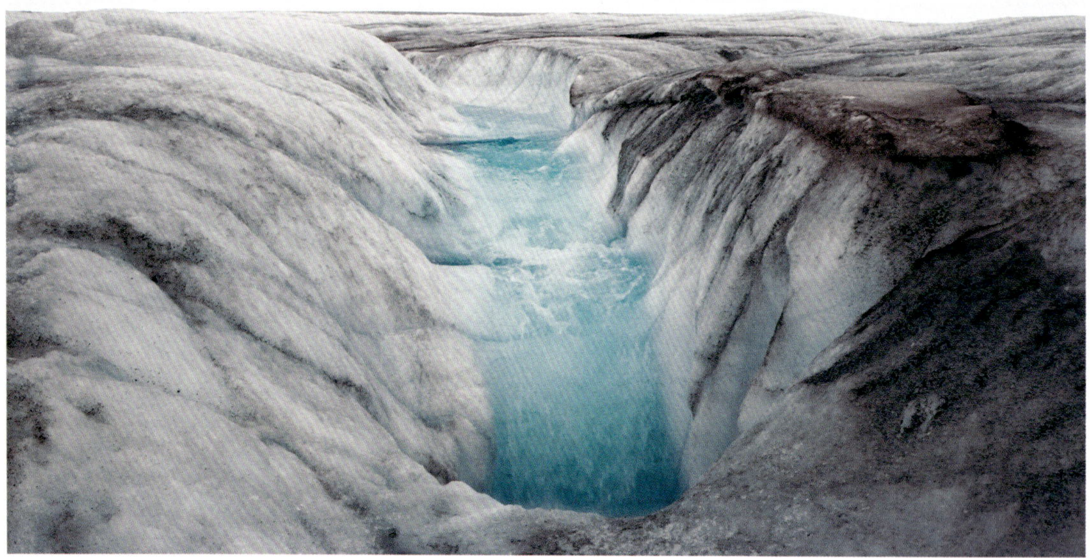

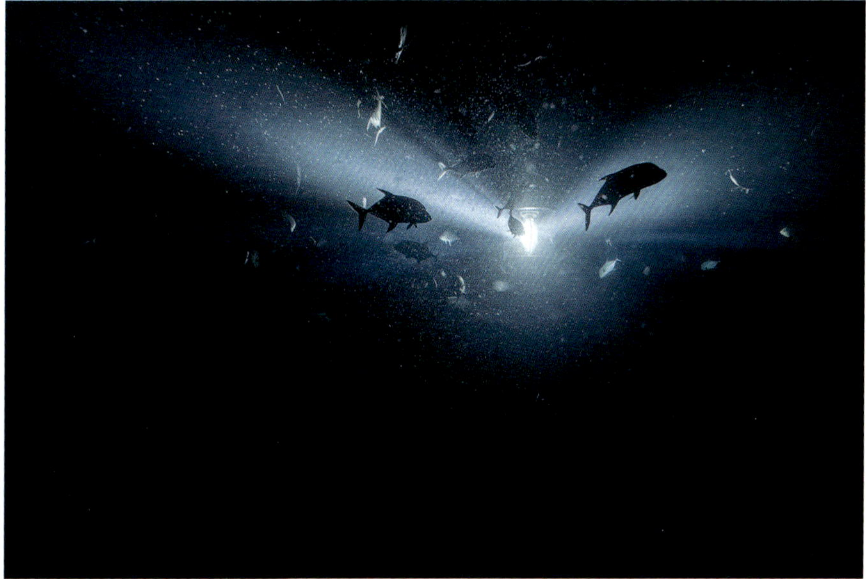

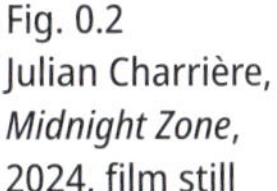
Fig. 0.2
Julian Charrière,
Midnight Zone,
2024, film still

Can the Seas Survive Us? explores the diverse forces that influence humanity's interactions with aquatic ecosystems, marine biodiversity and the complex web of life within our water bodies. Through a series of exhibitions – including *A World of Water*, *Darwin in Paradise Camp: Yuki Kihara* and *Sea Inside* – projects to engage communities across East Anglia and the present publication, we hope to showcase the pivotal role that art and culture can play in reshaping our understanding of the seas. The essays that follow compel us to view nature not merely as a resource to be endlessly consumed and destroyed, but as a shared living habitat upon which all life ebbs and flows. They encourage a fundamental shift in our thinking, from extractive, exploitative systems towards models of coexistence and reciprocity, where we engage with nature in a more personal, regenerative and sustainable way. And they ask us to pause, to take a step back from the idea of the ocean as a sublime, unknowable vastness, and to instead evaluate more intimate and overlooked connections between humans and the seas, enjoying, as Pandora Syperek and Sarah Wade write in their chapter 'Sea Inside: Art and Marine Interiority' (p. 116), a 'more immersive view' that enables 'myriad aesthetic encounters, which may not be grand or awesome, yet in their varied intimacies offer important underexplored dimensions of these vital saltwater entities and the life that dwells beneath their surfaces'.

With sea temperatures climbing to unprecedented heights, the urgency of our environmental predicament is stark. This publication serves as a call to action, raising awareness of humanity's impact on marine environments and stressing

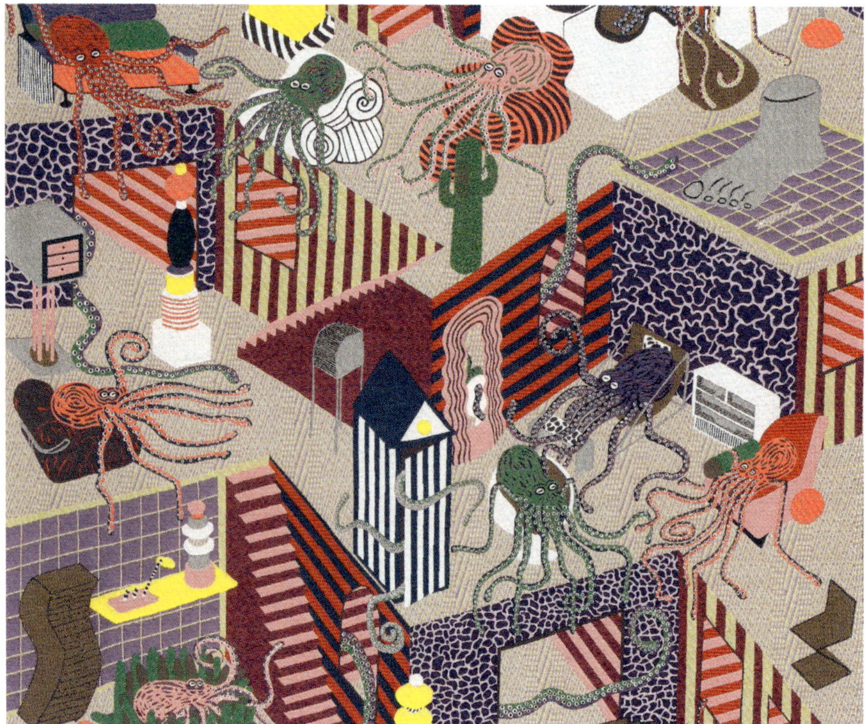

Fig. 0.3
Koen Taselaar,
Radical Furniture for Radical Times (detail),
2019, tapestry

the need for a collective response to the challenges posed by the climate crisis. It examines artistic responses to the multifaceted nature of water – both as a salient element for all life and as a symbol of the broader environmental struggle.

Despite the fundamental importance of water, it is too often regarded solely as a resource or capital. What is needed is a more nuanced understanding of hydrological cycles, one that recognises the intimate connections between human and water bodies. By reimagining that relationship on both a planetary and local scale, artistic and scientific collaboration, alongside the expanding discourse of the 'blue humanities' – a field that probes the cultural, historical and philosophical aspects of our relationship with water – can help us to gain a deeper insight into the ecological challenges we face. This interdisciplinary approach helps to illuminate the often-destructive relationship between humans and the seas, whilst fostering a renewed sense of responsibility for safeguarding the Earth's global commons.

The aim of this publication is to cultivate a deeper, more interdisciplinary understanding of how art, science, education, architecture, heritage and culture can inform and inspire new ways of thinking about our relationship with the natural world and our bodies of water. Specifically, it asks, as the world continues to warm and rising tides threaten coastlines, how do we rethink our connection to the sea? Can we create new models of engagement that promote harmony between human activity and marine ecosystems? Should we regard the sea as a unifying force rather than a divider of peoples and nations? And how can art and culture play a transformative role in addressing the pressing water-related issues of the climate crisis?

A WORLD OF WATER: SHAPING THE PAST, PRESENT AND FUTURE

> *Even if you never have the chance to see or touch the ocean, the ocean touches you with every breath you take, every drop of water you drink, every bite you consume. Everyone, everywhere, is inextricably connected to and utterly dependent upon the existence of the sea ... If you think the ocean isn't important, imagine Earth without it. Mars comes to mind. No ocean, no life-support system.*[6]
>
> Sylvia Earle, Oceanographer

Sylvia Earle's words encapsulate the profound significance of the oceans. Amid a climate crisis demanding global attention, it is paramount to understand the intricate connections between land, water and life. The ocean, our planet's life-support system, is in peril. Our failure to safeguard it endangers not only its equilibrium but the very existence of countless species. Addressing these challenges necessitates a fundamental transformation – from isolated efforts to a holistic, global perspective that recognises our deep entanglement with the seas.

The world's seas are intrinsically linked to life on Earth. They generate nearly half the oxygen we breathe, regulate our climate, drive the water cycle and absorb a substantial portion of humanity's carbon emissions. More than this, they harbour rich biodiversity, crucial migratory routes for whales, sharks, fish and birds, as well as extraordinary ecosystems like deep-sea coral forests and hydrothermal vents. Human activities have driven those ecosystems to the brink. In his chapter, 'Sounding the Unknown: Waters and Global Change' (p. 33), Soren Brothers uses bleached corals to viscerally demonstrate the sensitivity of our oceans and their biodiversity to climate change, emphasising the visible stresses that marine lifeforms are under as a result of our actions.

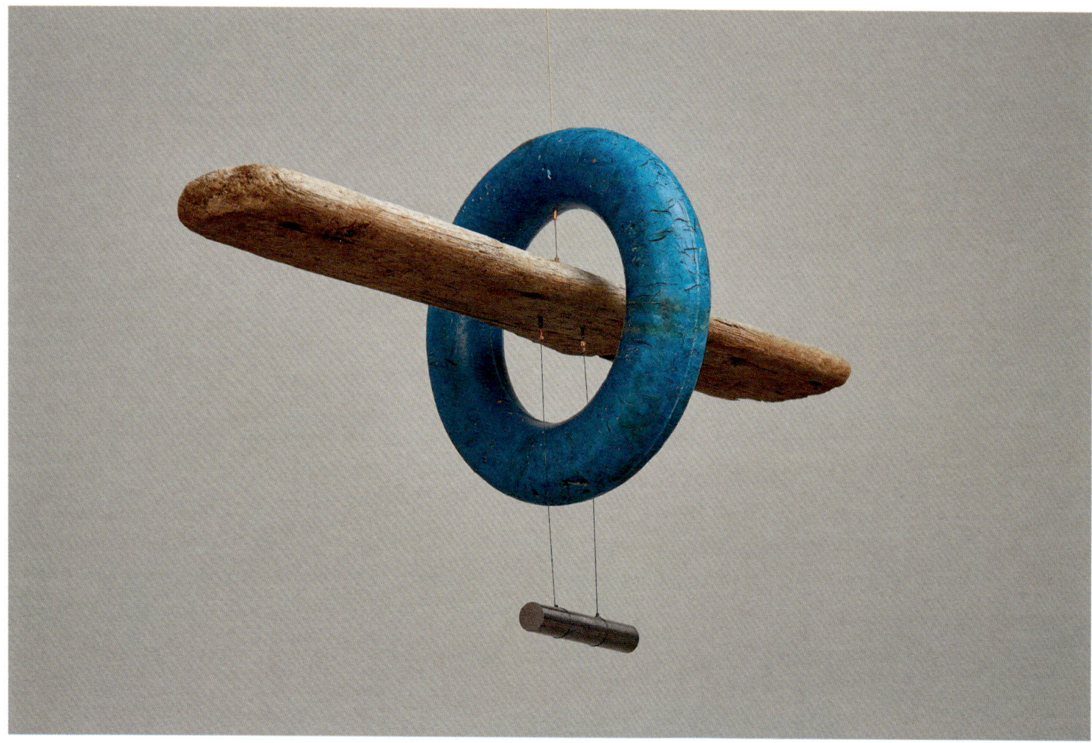

Fig. 0.4
Olafur Eliasson,
Shore compass (02:00, blue), 2018, driftwood, plastic, magnet and wire

From oil extraction in the North Sea to the devastating effects of nuclear testing in the Pacific and the mobilisation of natural resources during global conflicts, we have wrought havoc upon marine environments. Indeed, in Karen Jacobs's chapter, 'A Sea of Resilience: Oceanic (In)visibility' (p. 80), we see all too clearly the environmental and social impact of nuclear testing in the Pacific and the ongoing repercussions for the communities in the surrounding atolls. Dovetailing with this, Tania Moore's interview with Yuki Kihara, a Japanese-Sāmoan interdisciplinary artist, highlights the devastating result of climate change on Pacific communities (p. 95).

The pressing question is whether we possess the wisdom to save ourselves and our oceans. The recent surge in art exhibitions focused on ecological themes underscores the escalating urgency to confront the climate crisis. Art, with its power to evoke emotions, provoke thoughts and inspire actions, plays a crucial role in raising awareness and fostering dialogue about the most critical issues of our time. As Courtney Traub observes in her chapter, 'The Image of Eternity' (p. 56), we are seeing a growing number of contemporary works of art 'that attempt to grapple with phenomena such as oil spills; with waters and ocean floors permeated with microplastics; or with the threatened swallowing of island nations'. Similarly, in her chapter 'Tipping Points: Pausing the Unstoppable in the Work of Maggi Hambling, Claire Cansick and Margaret Mellis' (p. 68), Antonia Blocker delves into how these three artists use depictions of the sea as a metaphor for environmental tipping points, linking their work to broader concerns about climate change (fig. 0.5).

Rising temperatures do more than make the world hotter. Widespread weather impacts are one reason why scientists prefer the term 'climate change' to 'global warming'. It can be challenging to grasp just how intimately connected the climate and the water cycle are and to untangle the complex web of interactions between them. The water cycle, a fundamental process in the Earth's climate system, involves the continuous movement of water between the atmosphere, land and oceans. Any changes to it – such as greenhouse gases trapping more heat in the atmosphere – have broad, rippling consequences. The upshot is that phrases like 'the climate crisis is a water crisis' might sound like slogans, but they are fundamentally true. The IPCC describes rising temperatures as intensifying the water cycle, heightening its extremes and making the swings between them more erratic. Storms and rainfall become heavier and more destructive; droughts last longer and are more severe. Countries accustomed to one type of disaster must grapple with others; key regional weather patterns, like the monsoons of South Asia, become increasingly unpredictable.

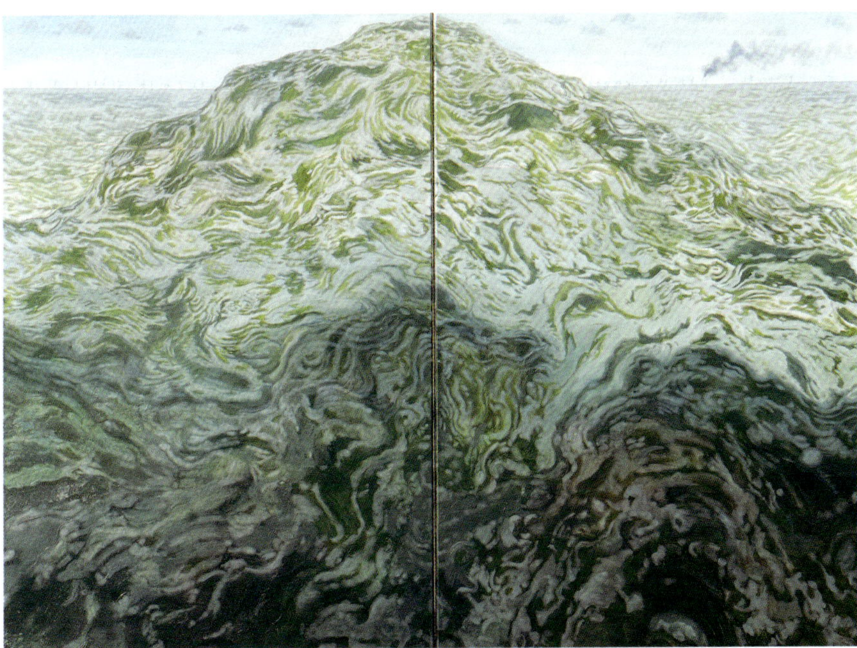

Fig. 0.5
Claire Cansick,
Scroby Sands, 2024,
oil on canvas

The consequences are severe. Rising sea levels, driven by melting ice and the expansion of warming ocean water, render freshwater sources saline, inundate coastal regions and propagate waterborne diseases. Flooding disrupts sewage and drinking systems, spreading pathogens that replicate faster due to warmer temperatures. As agriculture demands more irrigation, pressure on freshwater supplies increases, whilst hydroelectric dams struggle amid declining water levels. These disruptions force vulnerable communities towards unsustainable water sources, exacerbating the crisis.

OCEANS OR THE SEAS

In marine science, there is a tendency to refer to the world's waters collectively as 'the ocean', emphasising their interconnectedness. The ocean functions as a singular global system, circulating like an immense conveyor belt that regulates climates, supports ecosystems and links continents. Yet climate change threatens this system, slowing its natural rhythms and precipitating ecological upheaval. While this terminology is scientifically accurate, the everyday and cultural use of the plural 'seas' more aptly captures the diversity and individuality of each body of water.

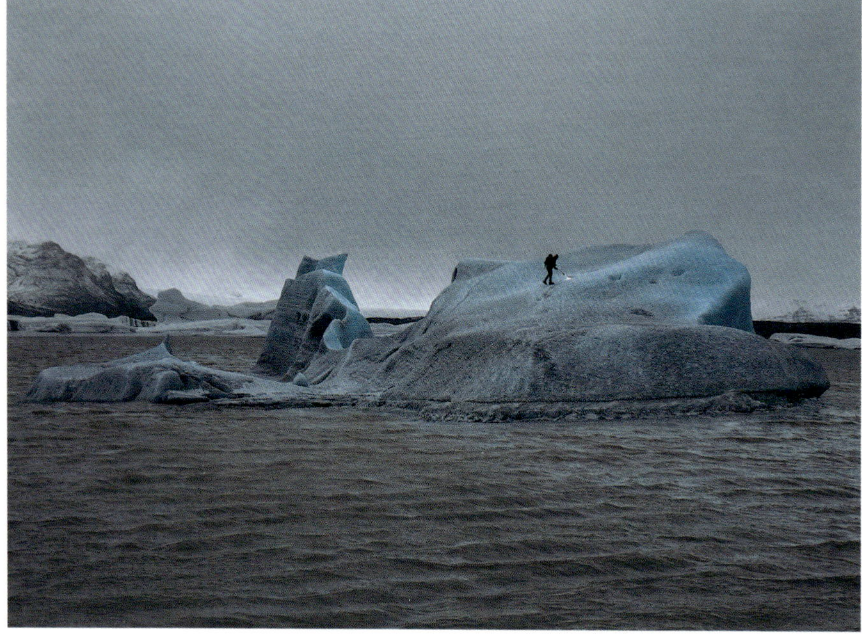

Fig. 0.6
Julian Charrière,
The Blue Fossil Entropic Stories I, 2013, C-print

Many people tend to use the terms 'sea' and 'ocean' interchangeably; however, from a geographical standpoint the two terms have distinct meanings. Seas are generally smaller than oceans and are located where the ocean meets land. They are often partially enclosed by landmasses. The ocean, therefore, represents a much larger body of open water than a sea.

In curating *Can the Seas Survive Us?* we deliberately chose the plural 'seas' to highlight the distinctiveness of each marine realm. Every sea possesses its own legal frameworks, characteristics, ecosystems and environmental challenges. In her chapter, 'Fluid Lines: Charting Change and Changing Charts' (p. 25), Djoeke van Netten considers how the ever-shifting nature of coastlines defy traditional mapping techniques while also challenging the motives behind cartographic endeavours, with their desire to 'make sense of, to structure and to control the Earth', while 'failing to represent some of the most remarkable features of the real seas'.

Protecting individual seas, where local actions yield tangible effects, is more conceivable than the seemingly insurmountable task of saving the singular 'ocean'. In his chapter, 'Estuarine' (p. 105), Patrick Flores shifts our perspective from the global to the local, in this instance from the vastness of the sea to the estuary, a liminal space where land and water entwine. Here, the flow of tides weave between islands and cities, threading through archipelagos. The estuary offers a more intricate lens on the sea, enriching the climate crisis narrative with its subtle rhythms, intermixing ethnographic insight with aesthetic nuance.

In Norwich, we have the opportunity to take meaningful steps to safeguard the North Sea by integrating interdisciplinary approaches across art, design, architecture, heritage and material culture. Such collaborations can set a precedent for how local efforts might inspire broader change.

SAILING THE NORTH SEA: A RESEARCH METHODOLOGY

In *The Edge of the World: How the North Sea Made Us Who We Are* (2015), historian Michael Pye contends that the North Sea has long been a crucible of technological, cultural and economic innovation in northern Europe.[7] He illustrates how this 'cold, grey sea' shaped modern civilisation, highlighting the dynamic interactions across the land–sea boundary. However, centuries of industrialisation and resource extraction have dramatically altered its landscape.

Today, the North Sea is a complex network of international maritime laws, energy infrastructures and shipping routes, turning it into a highly engineered, urbanised expanse. Offshore wind turbines, located far out to sea to minimise their impact on people and preserve coastal views, prompt questions about how much shoreline is needed to meet global energy demands. Offshore wind is one of the most expensive forms of renewable energy and wildlife campaigners also raise concerns about the risks these turbines pose to birds and the disruptive effects their underwater structures have on marine ecosystems.

Our curatorial approach for *Can the Seas Survive Us?* involved reimagining the exhibition as a space for both theoretical and practical exploration – exhilarating and hopeful. We sought to balance research, art, collaborations and narratives to provide a nuanced understanding of the challenges facing the North Sea and other waters globally. Addressing these issues required interdisciplinary cooperation, where art, design, heritage, culture, science, policy and lived experience converged to envision a future in which humanity coexists synergistically with the Earth.

Recognising that the seas cannot be fully understood through a single discipline or national lens, we embarked on a research voyage. The seas require long-term historical and geographical comprehension, calling for new tools, perspectives and methodologies. On 4 September 2024, under a radiant sky, we set sail from Lowestoft, England, bound for Rotterdam in the Netherlands. The North Sea, with temperatures ranging from 13 to 20 degrees Celsius and gentle, steady winds, seemed to welcome our expedition. Aboard *Excelsior* LT472 – a historic fishing

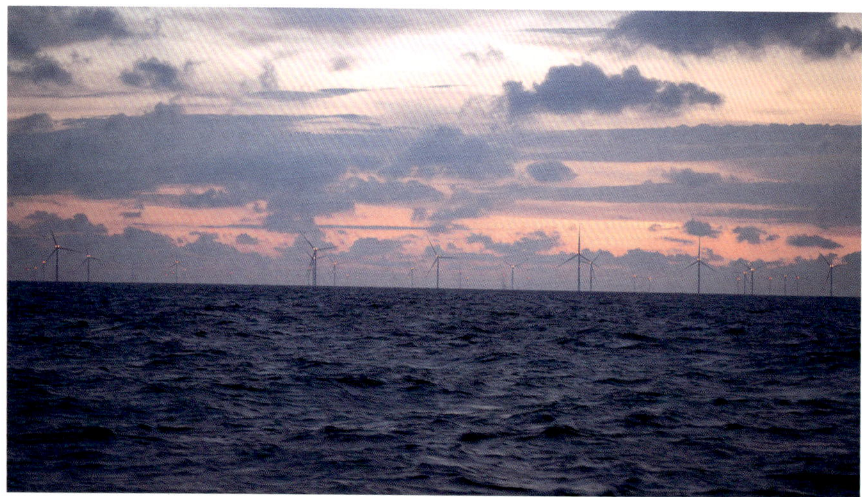

Fig. 0.7
Arieh Frosh and Ed Compson, *North Sea Night Turbines* (Windpark Hollandse Kust Zuid), 2024, film still. Norwich University of the Arts

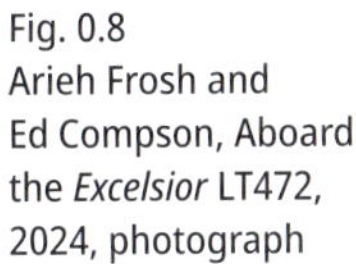

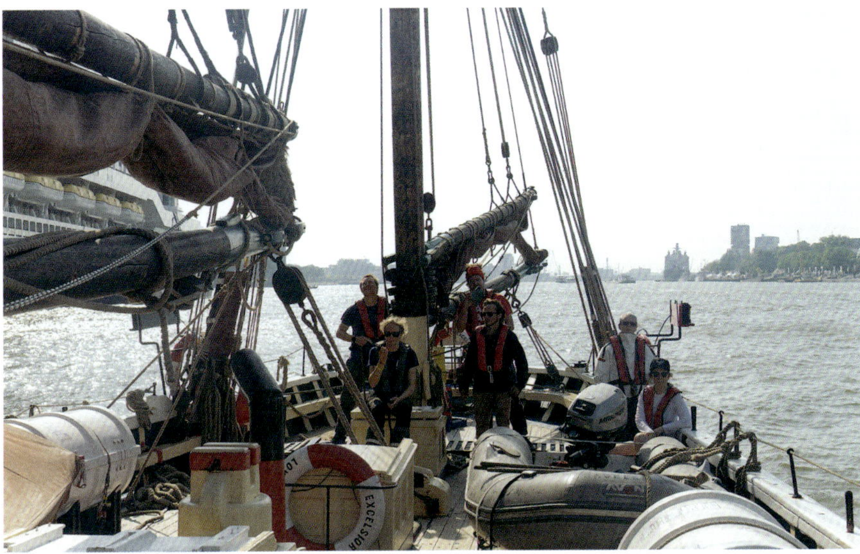

Fig. 0.8
Arieh Frosh and Ed Compson, Aboard the *Excelsior* LT472, 2024, photograph

smack built in 1921 and one of the last surviving Lowestoft smacks – our journey was more than a simple crossing. This ship, steeped in legacy, was not just a vessel but a symbol of the passage of time and the ever-changing seas that have witnessed centuries of human endeavour. Her 23.5-metre frame, with a capacity of 17 passengers, embodied resilience. This voyage continued from a 35-mile walk from Norwich to Great Yarmouth – a slow pilgrimage inviting reflection on climate action, deep time and fostering meaningful connections about our shared responsibility for the fragile ecosystems that sustain life (fig. 0.8).

Our crew – a diverse and intergenerational collective of artists, curators, archaeologists, educators, cultural workers, activists and academics from institutions such as the Sainsbury Centre and the Tyndall Centre at the University of East Anglia – included Jago Cooper, Rauri McGibbons, Erik Hartin and myself. Colleagues from Norwich University of the Arts – Louis Nixon, Candice Allison, Teresa Stoppani, Arieh Frosh and Ed Compson – joined us, along with Kaavous Clayton and Jules Devonshire from Original Projects in Great Yarmouth.

Led by our skipper, Charlotte Hathaway, and her capable crew – Ben Mousseaux, Harry Morfitt, William Lund and Lucas Owens – we sailed as one. Our conversations rose and fell with the gentle swells of the sea, echoing the natural rhythm around us. The waves lapping against the hull seemed to mirror our thoughts, as though the sea itself were part of our dialogue, its eternal pulse intertwined with our reflections.

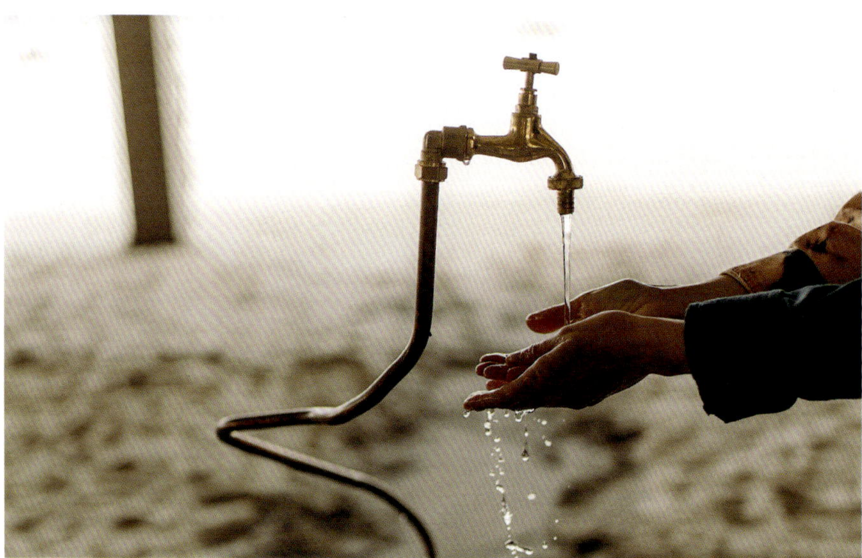

Fig. 0.9
De Onkruidenier,
Spiral Fosset, 2024,
metal, copper tubes
and sand

Beneath the tranquil surface of our voyage lay a profound question: *Can the Seas Survive Us?* These waters have nurtured life and sustained human civilisation for millennia, yet today they bear the weight of our unchecked ambitions. Once thought boundless, the oceans have become thoroughfares for commerce and repositories for waste. We have pushed them to their breaking point.

CONFRONTING THE CLIMATE CRISIS THROUGH ART AND RADICAL COLLABORATION

Our voyage was a philosophical and artistic exploration of exigent questions. How can we foster the trust and partnerships necessary to address the greatest existential challenge of our time? How can these ideas manifest within an art exhibition? What role can educational and cultural institutions play in accelerating climate action? How can art, design, architecture, museums and activism unite to influence policymakers and drive the green transition? Surrounded by the expansive sea and sky, we contemplated how art might offer a vision of a more sustainable future – one where humanity exists in cohesion with the planet.

As we prepared to set sail from the Maritiem Museum in Rotterdam on 9 September 2024, bound for Lowestoft, our journey swiftly shifted from adventure to adversity. The unexpected closure of the Erasmus Bridge, a key route out of the city, delayed our departure. The following day, the North Sea

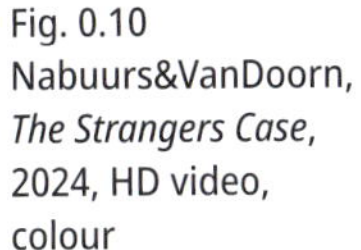

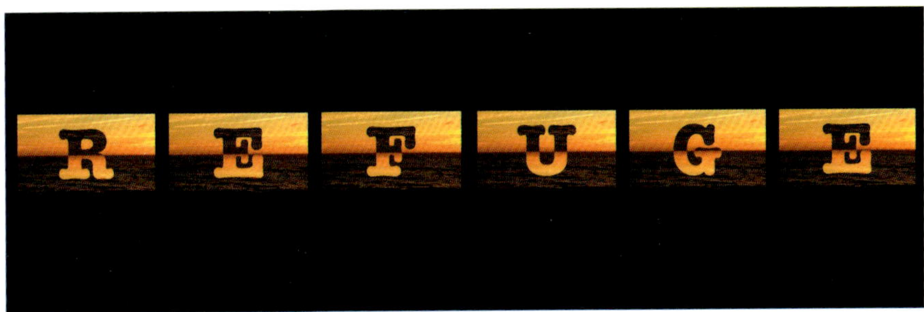

Fig. 0.10 Nabuurs&VanDoorn, *The Strangers Case*, 2024, HD video, colour

reasserted its dominance. Gale-force winds whipped across the waters, and a brewing storm rendered the voyage impossible for *Excelsior*, our century-old vessel. Reluctantly, we abandoned ship, dispersing across Eurostar trains and ferries – our return dictated by the seas' unpredictable temperament.

This moment served as a stark reminder of the sea's raw, untameable power – a force that remains beyond human control, the ultimate arbiter of our fate, indifferent to our progress. It is a humbling lesson: for all our human ingenuity – bending rivers, building ships, creating polders and reclaiming land – nature still dictates the terms of our survival.

CONCLUSION: REVIVING OUR SEAS REQUIRES A CLEAR-EYED REALISM AND BOUNDLESS IMAGINATION

Reflecting on our research journey one undeniable truth emerges: whilst the ocean's vastness may seem eternal, the life it sustains is anything but invulnerable. The degradation of our seas, primarily driven by human activity, stands as one of the most crucial and solvable crises of our time. Yet solving it demands far more than technological advances; it calls for coordinated global action and a collective will to act.

The path to meaningful change begins locally, particularly in fragile coastal ecosystems – the vulnerable spaces where land meets the sea. In this vein, Andrew Watkinson's chapter, 'Redrawing the Line' (p. 43), focuses on the constantly changing coastlines of East Anglia, highlighting how rising sea levels and coastal erosion reshape the landscape. Combining scientific data, historical narratives and art history, Watkinson underscores the fragility of our shores to changing sea levels and the delicate balance between human activity and environmental change.

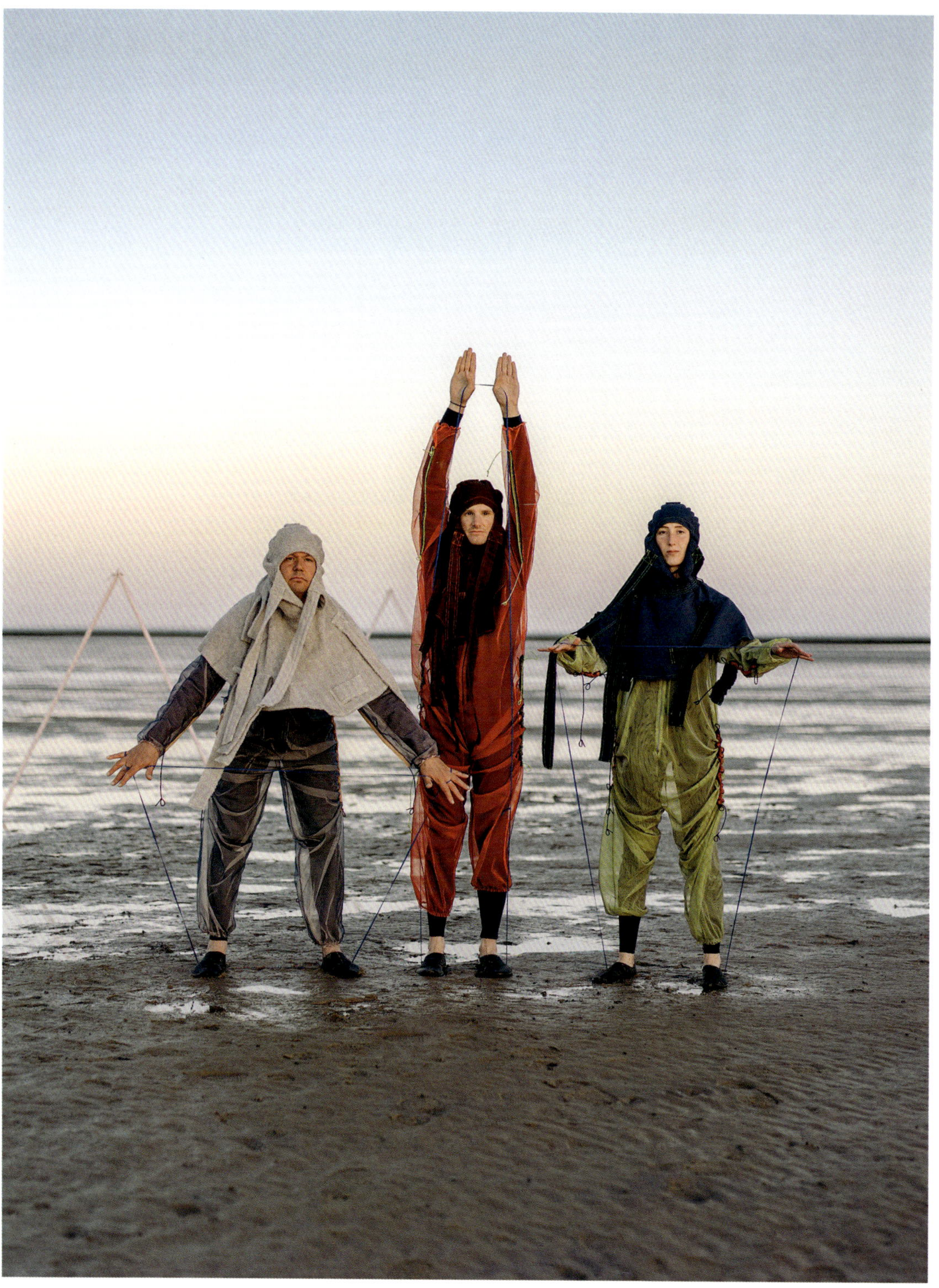

Fig. 0.11
De Onkruidenier, *Relearning Aquatic Evolution*, 2024, photograph

These coastal areas are already under immense pressure, but the broader challenge is global. The health of the world's seas is a shared responsibility requiring bold international cooperation. The seas possess an extraordinary capacity for regeneration, capable of restoring their biodiversity and abundance, but this potential can only be realised if we fundamentally rethink our relationship with nature and recognise, with humility, the limits of our planet's resilience.

Inspired by *Designing Climate Solutions: A Policy Guide for Low-Carbon Energy* by Hal Harvey, Robbie Orvis and Jeffrey Rissman, an effective climate strategy must strike a balance between clear objectives and practical methods for achieving them.[8] Addressing climate change cannot rely on vague promises, superficial solutions or greenwashing. The focus must be on impactful actions that shift human behaviour, reshape societal values and challenge the capitalist and consumerist systems that drive overconsumption and accelerate nature loss.

Can the Seas Survive Us? underscores the indispensable role of the ocean in sustaining life and presents a compelling invitation to act. The book investigates humanity's deep and often destructive relationship with the seas through a diverse constellation of artistic, scientific and critical perspectives. Interwoven throughout the book are a series of sea shanties by Harun Morrison, beginning with 'The Leaving of Kattegat Sea (or The Mermen's Lament)' and concluding with 'The Ghost Whales'. These poems reflect on themes of freedom, possession, and the tension between land and sea, offering a lyrical exploration of humanity's fraught relationship with the ocean.

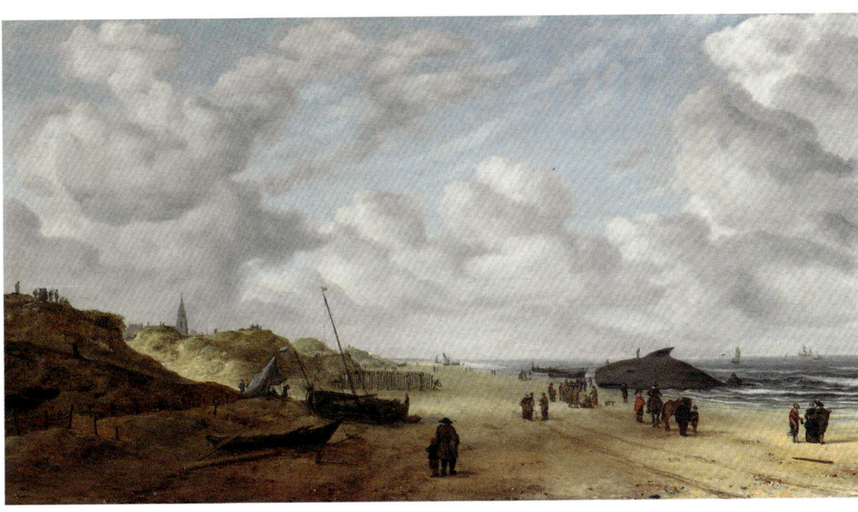

Fig. 0.12
Hendrick van Anthonissen, *View of Scheveningen Sands with Stranded Sperm Whale*, *c.*1641, oil on canvas. The Fitzwilliam Museum, University of Cambridge

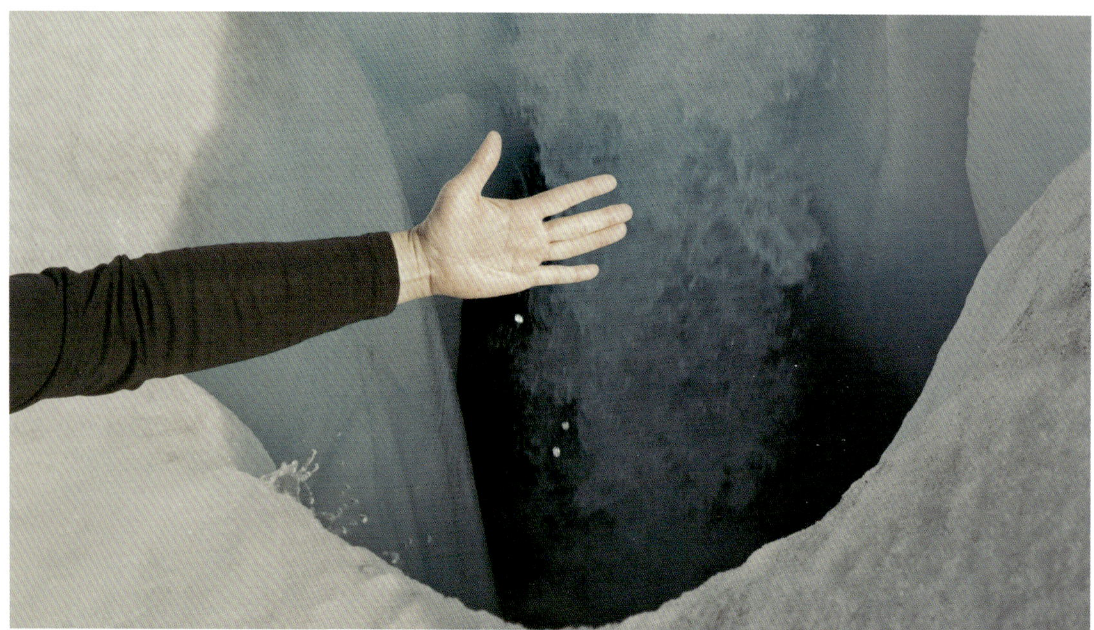

Fig. 0.13
Julian Charrière,
Pure Waste, 2021,
film still

By confronting these pressing issues, this book aims to inspire a sense of collective responsibility and catalyse high-impact action to safeguard our global commons for future generations. By identifying the root causes of climate change and understanding who holds the power to make imperative decisions about the future of our seas and oceans, this approach, whilst lacking the revolutionary fervour of a Marxist revolution, offers a pragmatic and effective way forward – one grounded in tangible action and lasting change. It challenges readers to rethink their relationship with the waters that sustain us, empowering art and culture to engage deeply with the multifaceted dimensions of climate change – cultural, scientific, emotional, philosophical and spiritual.

FLUID LINES: CHARTING CHANGE AND CHANGING CHARTS

DJOEKE VAN NETTEN

Open your phone, now open Google Maps and zoom out as far as possible. Some observations:

1) All oceans and seas are baby blue and there is no difference between, for example, the Arctic Ocean (partly frozen over) and the Mediterranean (almost 30°C last time I tested).

2) The borders between land and sea are represented by static, clear lines (fig. 1.1).

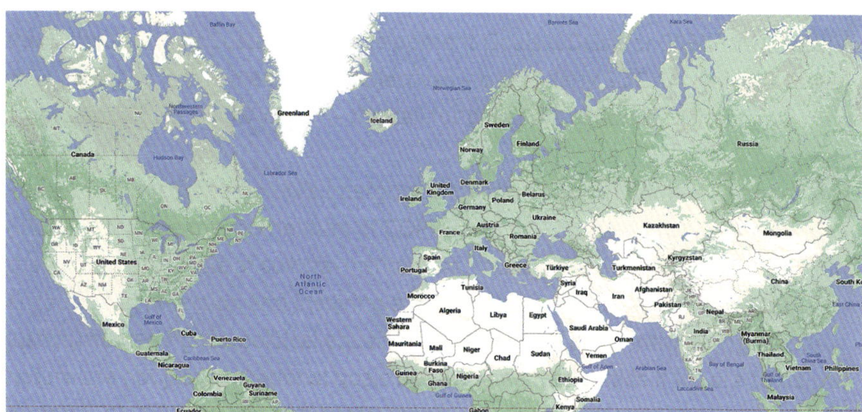

Fig. 1.1
Google Maps,
Map data, 2024

I wrote this essay from the beach; staring over my laptop into the North Sea, the coast of East Anglia invisible in the far distance.[1] The sea is 50 shades of dark blue, sometimes close to grey or green, here and there topped with white. There is no line, no fixed edge. The only continuity is change; a constantly moving boundary between land and water. I know about rising seas, and not just the twice daily back and forth of the tide, but an ever-nearing danger, threatening places like my home country of the Netherlands, which has over a quarter of its land below sea level.

Exploring the theme *Can the Seas Survive Us?*, this essay examines how modern and premodern maps struggle to capture the dynamic nature of coastlines and oceans. Those lines are fluid and ever-changing. Yet the boundary between land and sea is a fundamental dichotomy and has been so since the third day of Creation. And if you do not believe in Creation, it is nonetheless a cultural distinction apparent in creational stories the world over. This border is the one most pronounced on maps, and yet it is the one least discussed. The academic discipline of Border Studies usually examines borders between states. How,

I wonder, can we map seas and coastlines taking fluidity and variation into account? What can the history of cartography teach us in this respect?

Old maps appeal to the public imagination, both for their sheer visual beauty and for their content. But what do we see when we look at a map? While most people tend to assume a world map presents the real world, the key ingredient of every map is, in fact, imagination. Maps in general, and coastlines on maps in particular, never just show reality; they embody all kinds of assumptions, beliefs and technologies.[2] Coastlines are arguably one of the most difficult features to map since they are impossible to measure. Counterintuitively, coasts do not have a well-defined length, not only because of the constant movement mentioned above, but also since the result you get depends on your units of measurement and scale. This is known as the 'coastline paradox'. It may be helpful to engage in a thought experiment: how would outcomes differ when measuring Britain's coast using units of 100 kilometres, or 10 kilometres, or using the length of individual molecules?[3]

In the authoritative book series *The History of Cartography*, maps are defined as 'graphic representations that facilitate a spatial understanding of things, concepts, conditions, processes, or events in the human world'.[4] The usefulness of this academic, all-purpose concept can be debated, but the term 'representation' immediately indicates that a map is not the real thing; it is a model. Writers of fiction, like Lewis Carroll (1832–1898) and Umberto Eco (1932–2016), have played with the idea of a map on a 1:1 scale.[5] In practice, maps are much smaller, and consequently most features on Earth do not appear on maps, or they are rendered largely exaggerated. The smaller the scale, the more generalising cartography becomes, denying variety and individuality: every capital a red square, every other large city a red dot, every smaller place erased.[6] There are also many features on a map that are not on the Earth as such, like those red squares and dots, as well as lettering and (most) borderlines.

The largest issue in mapmaking has always been the spherical nature of the globe. How can we present a round object on a flat surface? Think of peeling an orange and being challenged to puzzle the peel into a neat square or two circles. The various mathematical solutions that have been proposed over the centuries are never neutral, never innocent, and demonstrate as well as determine world views. Google Maps is a case in point again. Even if other projections and orientations have been proposed – Australia-centred maps with South at the top, or the Gall–Peters projection promoted by UNESCO – we are too used to the already over 450-year-old Mercator projection to see beyond it.[7] It shows Europe, relatively large, on top and in the middle, all other continents in the periphery by definition, and the Global South relatively small when compared to its actual size in square kilometres.[8] Moreover, since the Pacific is parted at the edges, most world maps fail to show that over 70 per cent of the surface of the globe is covered with water.

The proportion of the Earth's surface that water took up on more historical maps was smaller still. Typical Medieval European maps are known as T-O maps, since they are circular with a watery T-shape dividing three continents (Asia, Europe and Africa). The surrounding O consists invariably of ocean, an all-encompassing body of water that, at least visually, has the same consistency and characteristics anywhere around. In theory, this ocean would facilitate global connections; in practice nobody sailed around it. Monsters lived there.

At the end of the thirteenth century, a list of places and their coordinates was graphically put onto what we would recognise as a world map by the second-century Egyptian scholar Claudius Ptolemy (*c.*100–*c.*170). Latitude and longitude offered a grid that structured and thus tamed natural space. Early Ptolemy world maps have almost invariably bright blue oceans, painted in expensive lapis lazuli. They display much more land than water. Even with the 'discovery' (from a European point of view) and insertion of America, the Ptolemaic map view did not fundamentally change after 1492 (fig. 1.2).

A completely different way of presenting the seas is found in portolan charts. First produced in the Mediterranean from the late Medieval period, nowadays they are hailed as surprisingly accurate in service of navigation. Coastlines on portolan charts consist of curved lists of place names that require you to turn your head (or the map) to read them all in sequence. The seas are not coloured, but they do portray compasses and compass lines, laying out another human grid over nature.

From the second half of the fifteenth century, printing became available in Europe to reproduce and spread knowledge. With this technology new ways of representing the seas emerged, since colour printing was not yet available to mimic the drawn and painted maps that had come before. To distinguish land from water, oceans were hatched or shaded. Some maps were coloured, but this was done by hand after the printing process. The coloured maps are the ones we see in exhibitions and coffee table books, but we have to realise that most early modern maps functioned in black and white. Sixteenth-century seas thus consisted mostly of a varied range of dots, waves, straight or curved lines.[9] This

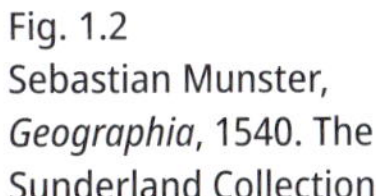

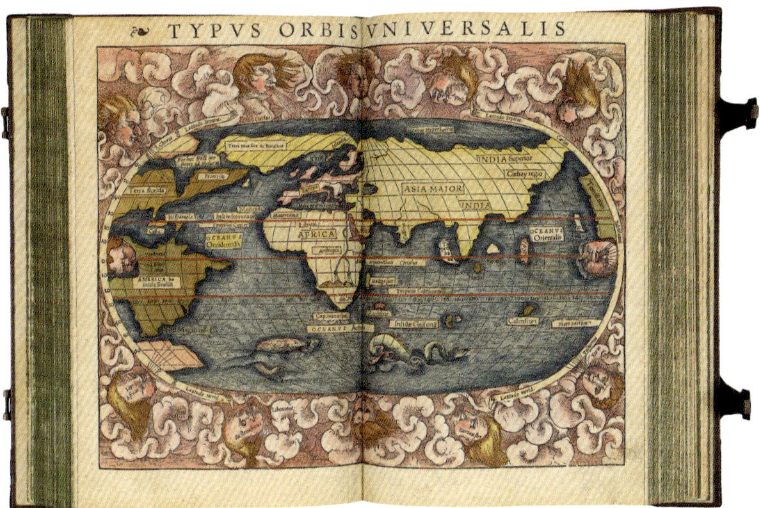

Fig. 1.2
Sebastian Munster, *Geographia*, 1540. The Sunderland Collection

shading of the sea happened using both wood blocks and (later) copperplate engraving to produce images. It also happened in printed world maps as well as in (larger scale) maps in printed navigational guides.[10] In any case, the filling was homogeneously distributed across the world, without recognising differences in depth, colour, temperature, tempests, etc. One exception is a world map by the Jesuit scholar Athanasius Kircher (1602–1680), who tried to map ocean currents, which he related to subterraneous volcanic activity (fig. 1.3).[11]

In around 1600 shaded seas disappeared almost completely from European printed maps with the suggestion that publishers economised for commercial reasons. Yet, despite this lack of shading, most seventeenth-century world maps did not present completely empty seas.[12] On the contrary, they displayed grids (showing longitude and latitude, compass lines, or both), compasses, lettering and ships. Especially since ships evermore replaced the sea monsters that adorned sixteenth-century oceans, it can be argued that mapmakers emphasised human inventions (in a ridiculously large scale, compared with, for example, islands) to demonstrate the domination of man over nature.[13] Other additions to world maps, which can be considered in the same vein, include the travel routes of famous 'discoverers' (always white men), crossing oceans, circumnavigating the world, connecting continents.[14]

While the shading of whole oceans disappeared, on many, especially small-scale, maps, the coast became shaded, though without discriminating between various coastal types.[15] The shading is mostly found on the outside of the land (while portolan maps had their texts forming coastlines inland), rendering the

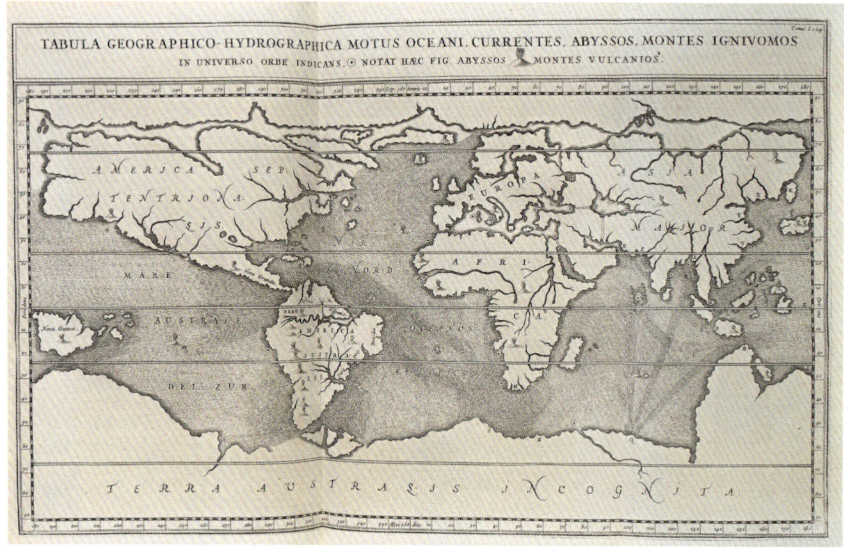

Fig. 1.3
Athanasius Kircher, *Mundus Subterraneus*, 1664–5. The Sunderland Collection

shallowest sea the darkest. This is an interesting contrast with world maps today, where the deepest parts of the ocean are coloured in darker tones (Google Earth and Apple Maps, among others, have this).

Pilot guides, books to learn the art of navigation, emerged in Holland in the 1580s. Lucas Jansz Waghenaer (*c.*1534–*c.*1606) was the first to compile such a book: *The Mariner's Mirror* (1588). This genre combined three different means of presenting coastlines. The most easily recognisable, and the most studied by historians of cartography, are charts. They view the Earth from above, rendering coastlines immovable though sometimes with a shaded, or otherwise darker, foreshore and numbers indicating depth. This is the first time the sea is more than a surface. In these, and other navigational charts, the sea, at least the sea close to land, is not completely homogeneous anymore, but features anchorages, buoys and wrecked ships (fig. 1.4).

The second type consists of coastal profiles, demonstrating the verticality of the coast from the sea and showing hills, cliffs and some prominent buildings like churches and windmills.[16] Curiously, the sea is completely absent in these images. Finally, pilot guides contain a lot of text. Mostly ignored by historians of cartography, these texts narrating routes were referred to by contemporaries as maps (in Dutch '*leescaerten*', that is, 'maps to read').[17] Moreover, contemporary sailors preferred using these written charts over their drawn counterparts.[18] These texts take change and variation into account, discussing tides, currents, waves and colour, as well as soil and some fauna and flora.

Navigational maps aimed to be as up to date as possible, since safe sailing was literally a matter of life and death. Texts and tables show hourly and daily changes, while longer-term changes of shallows and coastlines can be discerned in larger sequences of editions of maps and map books. These are practical guides, describing, but not concerned with explaining, tides or changing currents. As such, climate change is not considered. Before the nineteenth century the word 'climate' referred to a fixed zone of the globe with certain characteristics. The idea that climate can change is a modern invention. The climate did change, however, in the early modern era, a period overlapping with what is called the 'Little Ice Age'.[19] While their global cooling contrasts with our global warming, it also came with extreme weather, making it worthwhile to compare past and present representations and reactions.

Juxtaposing then and now, the danger men posed to oceans was not, historically, as worrisome as it is today. Destructive fishing, exploitation for oil, gas and minerals, nuclear testing, plastics, chemicals and acidifying were mostly beyond the horizon. Human greed using the seas, however, arguably started in the early modern period and cartography played a crucial role. Europeans who wanted to navigate to places to trade, colonise or plunder, used maps. Such maps were even considered state secrets.[20] Furthermore, maps functioned to show off the largeness and riches of empires, while commercial map sellers spread these messages to make money themselves.[21]

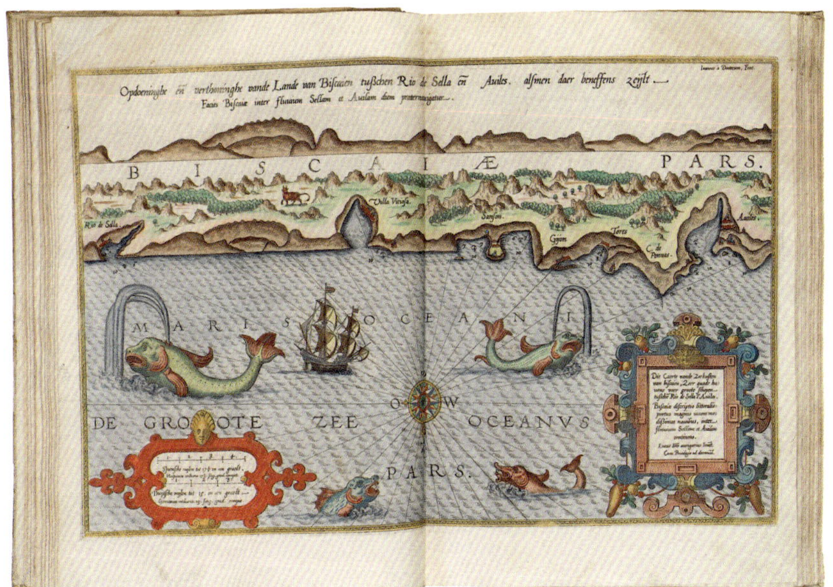

Fig. 1.4 Lucas Jansz Waghenaer, *The Mariner's Mirror*, 1588. The Sunderland Collection

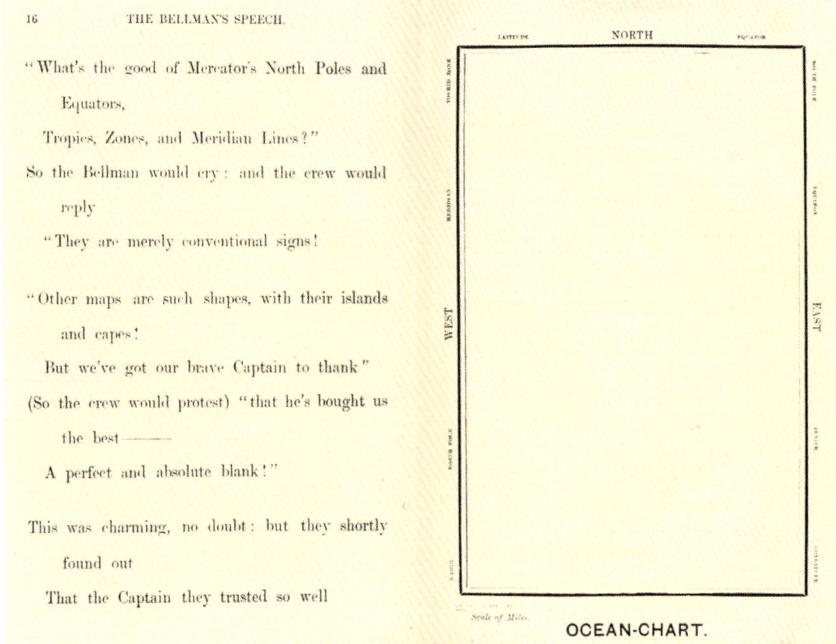
16 THE BELLMAN'S SPEECH.

"What's the good of Mercator's North Poles and
Equators,
Tropics, Zones, and Meridian Lines?"
So the Bellman would cry: and the crew would
reply
"They are merely conventional signs!

"Other maps are such shapes, with their islands
and capes!
But we've got our brave Captain to thank"
(So the crew would protest) "that he's bought us
the best——
A perfect and absolute blank!"

This was charming, no doubt: but they shortly
found out
That the Captain they trusted so well

Fig. 1.5
Lewis Carroll,
The Hunting of the Snark: An Agony in Eight Fits (Macmillan and Co., 1896).
The British Library, London

Thus, mapmaking has always been a human endeavour to make sense of, to structure and to control the Earth. Charting seas has almost always been charting coasts, connecting land and water, while failing to represent some of the most remarkable features of the real seas. It seems that only when using text is it possible to take fluidity and variety into account. However, over the course of the early modern period, navigational guides and atlases generally came with ever-less text. Besides, written maps are not much acknowledged in the history of cartography. What disappeared even more were monsters, ships and other decorative elements adorning the ocean on maps. This is beautifully illustrated by Lewis Carroll's ocean-chart showing a rectangular nothing (fig. 1.5).[22]

Hopefully a different ending is possible. Since our environmental crisis is also a crisis of the imagination, we should use that.[23] Maps can enable us to visualise the past, as well as possible futures. By showing the consequences of rising sea levels, we can imagine maps where my home country is more than half baby blue. There must also be more to the imagination than emptiness or one-dimensional baby blueness. The world's seas and oceans are not homogeneous. The challenge is to chart them as ever-changing and ever-evolving, as alive and endangered. The past does not provide ready-made lessons for the present, though it can serve as an inspiring inventory of possibilities.

SOUNDING THE UNKNOWN: WATERS AND GLOBAL CHANGE

SOREN BROTHERS

What is it about the ocean that draws us to it? Its sheer scale certainly has a gravitational pull on our imagination, but this seems to be an oversimplification somehow. Of course, we can rationalise its significance to us, and to all life on the planet, from the essential oxygen produced by marine algae to the role of oceanic currents in human history and climate futures. But beyond that, I'd argue that part of the ocean's pull is the romance of the unknown. When we look at the ocean we are typically not looking into it but across its surface, across an expansive reflection of the sky, and everything contained within it is contained within our imagination (fig. 2.1).

The otherworldly explorations of Jacques Cousteau (1910–1997) and deep-sea submarines provide only porthole-sized glimpses into a world which, for the most part, scientists try to make sense of through abstract numbers and figures. I understand the pull of this mystery not as a marine scientist, but as someone who has spent their professional career studying much smaller bodies of water; namely, lakes (fig. 2.2).

Fig. 2.1
Soren Brothers, Clouds on Gunnison Bay, Great Salt Lake, Utah, 25 March 2020, digital photograph

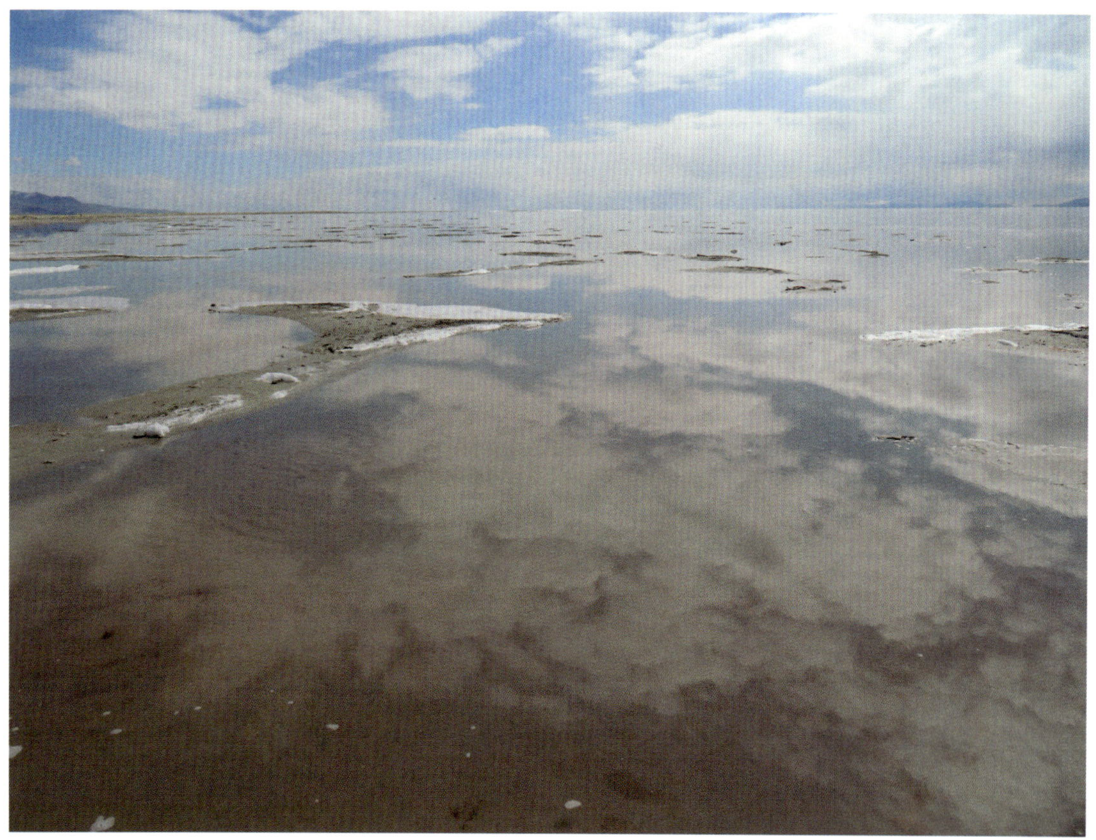

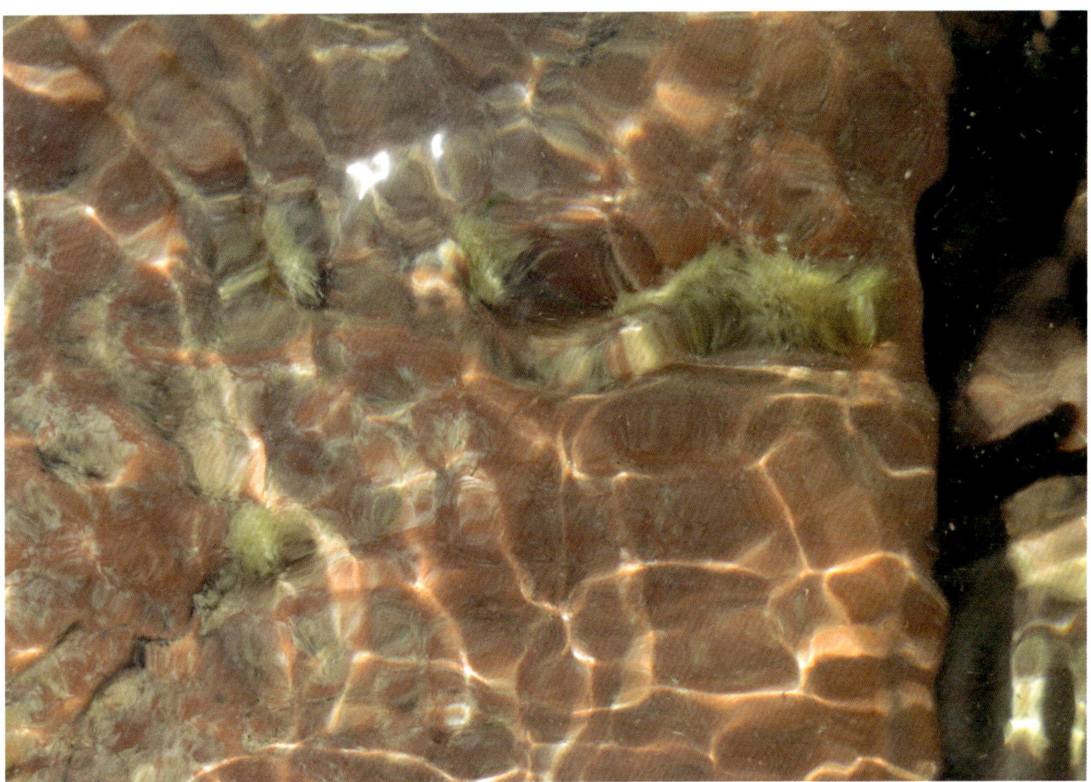

Fig. 2.2
Soren Brothers, Periphyton in Bear Lake, Utah, 7 September 2020, digital photograph

As a lake scientist (limnologist), I'm aware that scientific instruments provide us with data that form, in some sense, an abstract language. From our boats, we plunge equipment into the water and read out a code of numbers, revealed to us by the invisible depths below (fig. 2.3): '15.5 metres, 7.8 pH, 126 microSiemens per centimetre, 8.3 milligrams per litre D O, 120 per cent, 14.2 degrees, chlorophyll 12 micrograms per litre', and on and on.

The work that my colleagues and I do is ultimately an effort to translate that information, collected day after day for months, years, even decades, into a common language, as guided by our training in ecology, limnology, chemistry, physics, biology, geology and other fields. A lake can be perceived as a body or microcosm that integrates the changes that have been happening around it, whether to the atmosphere, land or waters, and expresses them through changes in its biological, geological, chemical and even physical characteristics. In other words, when we study a lake, we study the world. For this reason, lakes are considered 'sentinels of climate change', being amongst the best places to witness how our world is changing in ways both subtle and otherwise.[1] When an international group of scientists recently convened to determine whether the planet had entered a new geological epoch, the Anthropocene, defined by the changes that humans have made to planetary systems, the first step in the process was to locate the candidate 'Golden Spike' of the proposed epoch – the best place on the planet to showcase such a change. Perhaps not surprisingly, out of a shortlist of 12 sites around the world including caves, coral reefs and Antarctic ice cores, it was a lake – Crawford Lake, Canada – that was selected.

Indeed, lakes around the world are changing. Some changes are direct physical responses to a warming atmosphere – winter ice cover in cold and temperate-zone lakes is shortening and in places disappearing. Changes in ice cover can have far-reaching implications: in the high Canadian Arctic, longer annual periods of evaporative loss have resulted in the disappearance of some smaller lakes and ponds. Ice can have a profound impact on biological processes occurring within lakes by modulating the amount of light entering the water (thus controlling photosynthesis), and even by scouring rocks of algal accumulation from the previous year, providing an ecological reset each spring. Ice also acts as a barrier

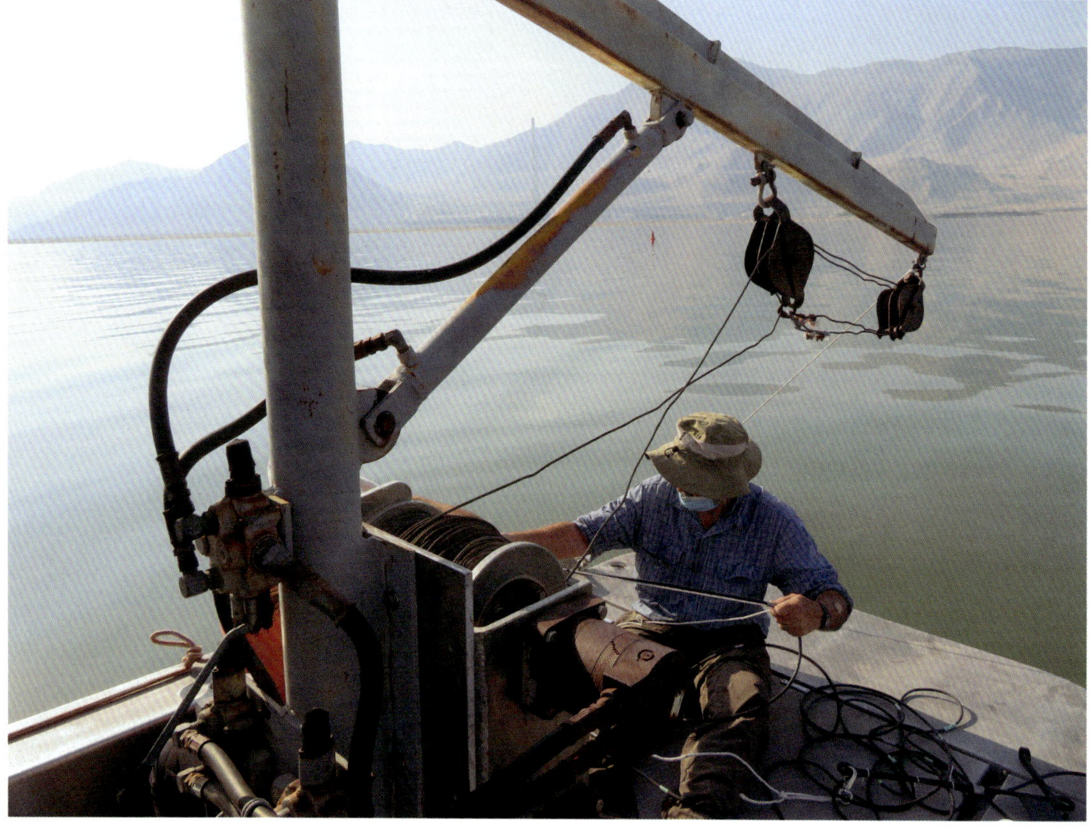

Fig. 2.3
Soren Brothers,
Field sampling on
Great Salt Lake, Utah,
21 September 2020,
digital photograph

to atmospheric oxygen entering lakes, which can in some cases lead to under-ice fish kill events if the organisms in the lake consume the dissolved oxygen before the spring melt. Changes in ice cover have even influenced the chemistry and biology of lakes to the point that their greenhouse gas emissions to the atmosphere have changed, signalling possible feedback mechanisms between our waters and climate change.[2]

Even in areas where ice is not part of the annual cycle of lakes, warmer air temperatures are producing warmer surface waters in lakes. Warming can reduce lake mixing, with the result that some lakes, such as North America's Lake Superior, are warming more rapidly than the local atmosphere. Warming lake waters can limit or eliminate fish habitats, fundamentally change microbial and algal dynamics (and consequentially, greenhouse gas fluxes) and even contribute to lower dissolved oxygen concentrations. The widespread and ongoing deoxygenation of lakes due to warming, reduced mixing and eutrophication, has been identified as a planetary boundary and key regulator of Earth systems stability.[3] All of this is to say that even the most fundamental physical responses of our lakes to climatic warming can easily spiral into food web and biogeochemical repercussions that can be difficult to predict. The future of lakes becomes significantly more complex when we add to that indirect climate impacts, such as changing patterns in precipitation or altered geographic ranges of aquatic species, and human influences, such as eutrophication and water withdrawals. However, lakes also provide us with understandings of how entire systems can fundamentally respond to disturbances, providing possible insights of broader value. In the early 1990s, Dutch limnologist Marten Scheffer (b.1958) and colleagues began to describe alternative equilibria in lakes subjected to increased nutrient loading.[4] Their work revealed that ecosystem changes are not necessarily passive or gradual – that a clear lake features biological feedbacks largely mediated by submerged plants, which keep it clear. However, at a critical point of turbidity, those submerged plants can no longer survive, and a new series of internal feedbacks kicks in, rapidly establishing the lake as a turbid, algae-dominated system with its own set of stabilising feedbacks making recovery efforts challenging. This work broadly expanded our understanding of critical transitions, or tipping points, in systems, which ultimately has been applied to ecosystems such as coral reefs and kelp forests, as well as social and planetary systems.

Needless to say, there is a lot we have yet to understand about lakes and how they are responding to, and interacting with, climate change. It is therefore understandable that the Global Ocean poses a kind of mystery that only pulls harder the more we dig into it. However, the small-scale processes described above for lakes all have analogues in how our oceans are responding to global change as well. Globally, the surface of the oceans has warmed by at least one degree Celsius over the past century.[5] Ice cover in the Arctic Ocean is reaching a record low, and new research is just beginning to show what this means for everything from food web dynamics to atmospheric greenhouse gas concentrations (with research indicating that the Arctic Ocean is likely currently acting as an expanding sink of atmospheric carbon dioxide). At the Southern Ocean, concerns have been raised that late sea ice formation may result in diminished algal content in winter ice, reducing winter food availability for krill, the base of the local food web.[6] Further from the poles, warmer ocean waters are an area of ongoing research, ranging from studies of mass coral bleaching events (occurring with just one to two degrees Celsius warming during heat waves, and which are many times more frequent today compared to just 40 years ago),[7] to less-publicised work exploring the physiological and community responses of phytoplankton to warming.[8] Although rising sea levels are often assumed to be solely the result of melting ice caps, the thermal expansion of our warmer oceans has also been a major driver of the approximately 20 centimetre rise since 1900.[9]

Just as the role of lakes as sinks or sources of greenhouse gases such as carbon dioxide is evolving, so is the role of the Global Ocean, which has absorbed roughly 40 per cent of all anthropogenic carbon dioxide emissions to date. This is resulting in ocean acidification, which is decreasing the capacity for the ocean to buffer our carbon dioxide emissions in the future, while also having far-reaching impacts on ocean biota. And just as lakes around the world are facing a crisis of deoxygenation, the volume of marine waters lacking oxygen ('anoxic') has quadrupled since 1960, particularly impacting coastal areas. Finally, recent work on the Global Ocean has also revealed tipping points, or critical transitions, that might occur with undeniably global consequences.[10] Foremost among these may be the albedo effect associated with the loss of Arctic sea ice – the presence of white snow and ice in the Arctic reflects solar radiation back to space, but as melting exposes more and more water, more radiation is absorbed by the ocean, leading to a rapidly accelerated warming feedback loop. Another worrying tipping point features the Atlantic Meridional Oceanic Current, driving the Atlantic Gulf Stream, which is slowing down and predicted by some to be nearing collapse. Should this happen, it is believed that a global loss of oceanic currents would follow. This would not only impact the global climate further, including by potentially increasing atmospheric greenhouse gas concentrations, but marine circulation may also play a role in preventing the accumulation of toxic gases associated with historic planetary mass extinction events.

As the challenges faced by the ocean present us with a global problem, we might expect that a global solution is necessary to tackle it. Some progress on this front has been made in recent years – while only 2.7 per cent of the Global Ocean area is formally fully protected, the recent High Seas Treaty, although yet to be fully ratified and implemented, is already considered an unprecedented legally binding success for marine conservation efforts, and a potentially instrumental platform for achieving goals of protecting 30 per cent of the world's oceans by 2030.[11] Still, focusing on the global scale, while valuable in grasping the scope of the issue, can be debilitating when it comes to taking individual steps towards its resolution. However, just as we are able to scale up from how climate change is impacting lakes to the parallel impacts at the scale of the Global Ocean, we can scale back down from the immensity of the global crisis to the value of small-scale and local solutions. In the Royal Ontario Museum, Toronto, where I work, we display bleached corals (fig. 2.4) that viscerally demonstrate the sensitivity of our oceans and their biodiversity to climate change. I inform visitors that 252 million years ago, the end-Permian ('Great Dying') mass extinction, in driving over 90 per cent of the planet's ocean life to extinction during a period of global warming, also eliminated the planet's corals, resulting in a five- to six-million-year gap in the fossil records before other organisms evolved to fill that niche. There is nothing any of us can do to prevent the next mass coral bleaching event from occurring. However, by eliminating the burning of fossil fuels and bringing global societies to net zero as rapidly as possible, by 2050 or sooner, we may reduce the frequency and severity of bleaching events occurring by the end of the century. Perhaps even more importantly for now, we can recognise that a coral bleaching event is not in itself the death of a coral, but rather a sign of stress, resulting from a dissociation between the coral polyp and the algae it lives symbiotically with. To this end, how we interact with our oceans and waters, including our fishing practices and water quality improvements, will influence the ability of corals to rapidly re-establish that symbiotic relationship, improving their chances to recover from bleaching events. In other words, rather than throwing up our hands and mourning the loss of marine life due to climate change, now *more than ever* is the time to double down on and improve sustainable practices of how we live with our lands and waters, becoming environmental stewards that play a supportive, rather than destructive or extractive, role with the natural world.

Whatever the reason for the ocean's pull on our attention and imagination, it is worth remembering that, through these interacting scales of influence and effects, we are intimately interconnected with its fate (fig. 2.5).

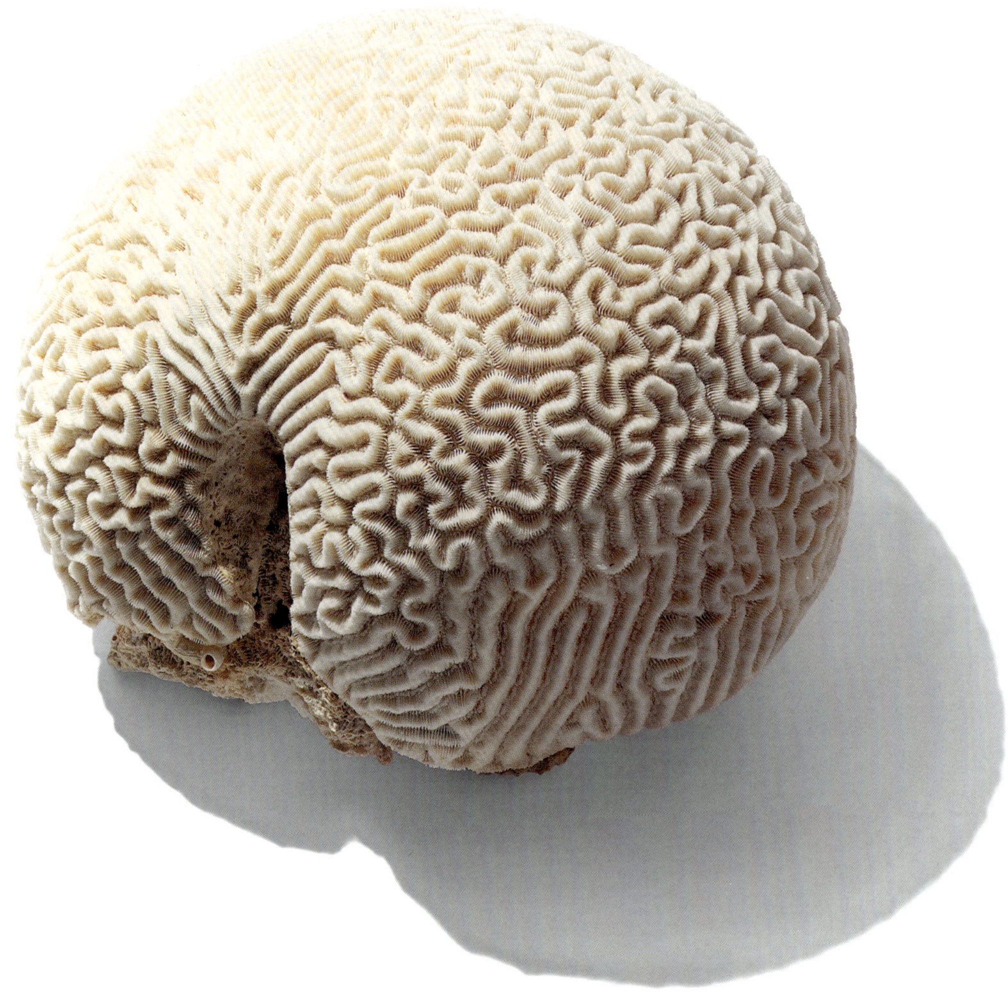

Fig. 2.4
Bleached brain coral. Royal Ontario Museum, Toronto, Canada

Our smallest individual actions collectively influence the ocean. An obvious example might be our decisions concerning what to eat on a given day – fisheries practices such as trawling have long been recognised as damaging to aquatic ecosystems, with trawling being banned in the River Thames by a 1394 English ordinance, and further continental restrictions on trawling gear being introduced in the fifteenth and sixteenth centuries. Another widely acknowledged connection is through our plastic waste that often ends up in the ocean, whether as macroplastics, which can kill birds and sea turtles that mistake it for food, or microplastics, the ultimate ramifications of which are still under investigation.[12] Our impacts on oceans also extend to our decisions on how to travel to work on a given day, considering the resulting implications to greenhouse gases and demands for critical mineral mining to support various modes of transport, as well as what clothes to purchase, given that the global textile industry is one of the foremost drivers of global change, with a hefty carbon footprint and outsized impacts on waste production as well as water quality and quantity. On the flip side, an improved stewardship of the Global Ocean may be a crucial component

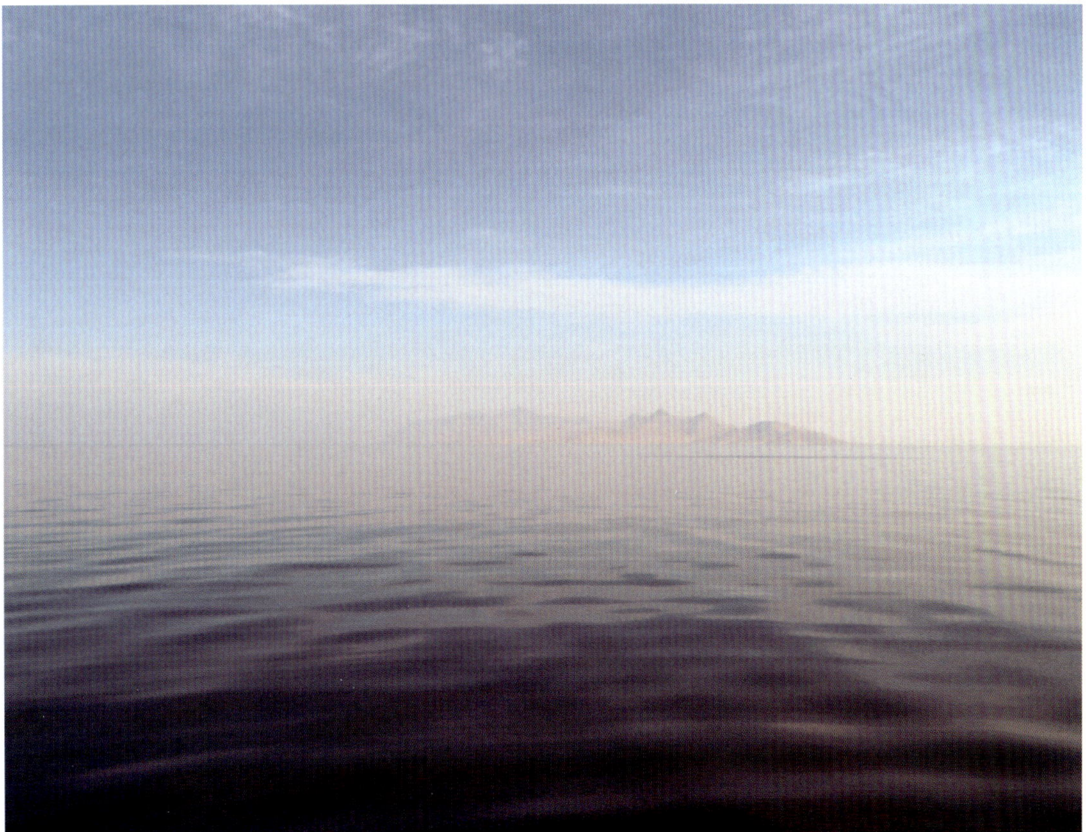

Fig. 2.5
Soren Brothers, Wildfire smoke on Great Salt Lake, Utah, 21 September 2020, digital photograph

of resolving the climate crisis, whether through expanding carbon-sequestering mangrove forests along its coasts or even recovering whale populations to prior levels, given the massive quantities of carbon that they individually sequester in the deep ocean, as well as the 'whale pump' process whereby they promote phytoplankton production, further drawing down carbon dioxide concentrations.[13] Ultimately, each of us, as individuals, regardless of our physical proximity or connection to oceans, benefit from the oxygen and climate stability that the ocean gifts us – and each of us stands to lose if those systems start to fail.

REDRAWING THE LINE
ANDREW WATKINSON

Cartographers draw a definitive line on the map giving a sense of permanence to our coastline and the borders of our nations. Yet one only has to look at historical atlases to see the impermanence of the borders of nation states, while early maps and the paleo record show how much the coastline has changed. The evolution of our coastlines over the last few hundred thousand years has, for the large part, resulted from changes in sea level, driven in turn by global temperatures and the amount of water and ice stored on land in a series of glaciations. Looking back to the last interglacial period, some 120,000 years ago, sea levels were between 5 and 10 metres higher than they are today, but then during the last glacial maximum, about 21,000 years ago, they were 125 to 134 metres lower, exposing large areas of land that that had previously been submerged below the sea; Great Britain was not an island.

It was following the subsequent melting and retreat of the glaciers that humans recolonised Britain. At the beginning of that period, much of what is now the North Sea would have been land, and you could walk from England to the Netherlands and the north of Denmark across an area now known as 'Doggerland'.[1] It was criss-crossed with rivers, forerunners of the Rhine, Meuse, Seine and Thames, and large grazing animals such as mammoths and woolly rhinoceros, together with bears, sabre-toothed cats and our early ancestors, roamed the land.[2] However, as sea levels continued to rise with the warming conditions, the coastline retreated, sometimes at a rate of 25 metres a year or more. Some 6,000 years ago, Doggerland disappeared beneath the waves and Great Britain became an island.

The overall shape of our coastline as we see it now was then visible, but it continued to change, especially in the east, as a result of rising sea levels and coastal erosion. There have also been land gains from accretion and land reclamation. The Broadland National Park in East Anglia is a particularly interesting case. As sea levels continued to rise, Broadland was initially inundated by the sea and became an estuary. In time, however, the movement of sediment by longshore drift along the coast led to the development of a sand bar across much of the estuary. This blocked the wide outlets of the estuary to the sea, deflecting the course of the rivers and limiting the amount of water that could enter them before the tide turned. Broadland habitats thus became dominated by freshwater, and estuarine conditions only persisted in the lower reaches of the rivers.[3]

This, however, was not to be permanent. Subsequently, around 2,000 years ago, gaps appeared in the sand bar and Broadland once again became an estuary.

It is the extent of the estuary as it was purported to be at the end of the first millennium that is depicted in the sixteenth-century 'Hutch Map' (fig. 3.1). The map is painted on the hide of a sheep, hence its shape, and is so named because it was stored in a hutch, a large chest, in Great Yarmouth town hall. The map shows how a number of the present-day settlements fringe the former estuary, and that the town of Yarmouth occupied a sand island at the mouth of the estuary; Norwich would have been a sea port.

In writing about the Hutch Map in 1826, the textile manufacturer and amateur geologist John Warden Robberds (1784–1871) wrote of the 'topographical inaccuracies, not to say absurdities' of the chart, but recognised that it demonstrated the long-held, if confused, notion that the valleys of Broadland had previously been filled by the North Sea, or as he referred to it, the 'German

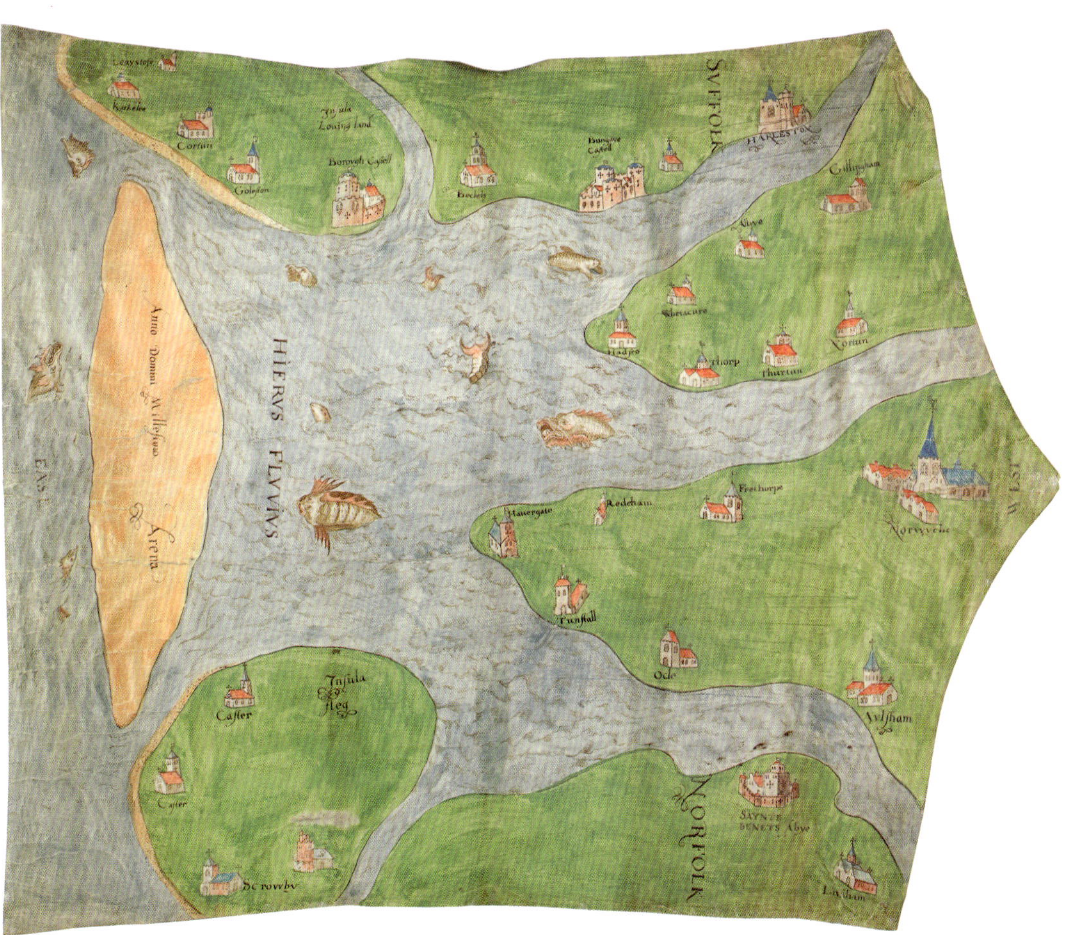

Fig. 3.1
'Hutch Map', sixteenth century, parchment membrane, drawn in ink and coloured. Norfolk Record Office

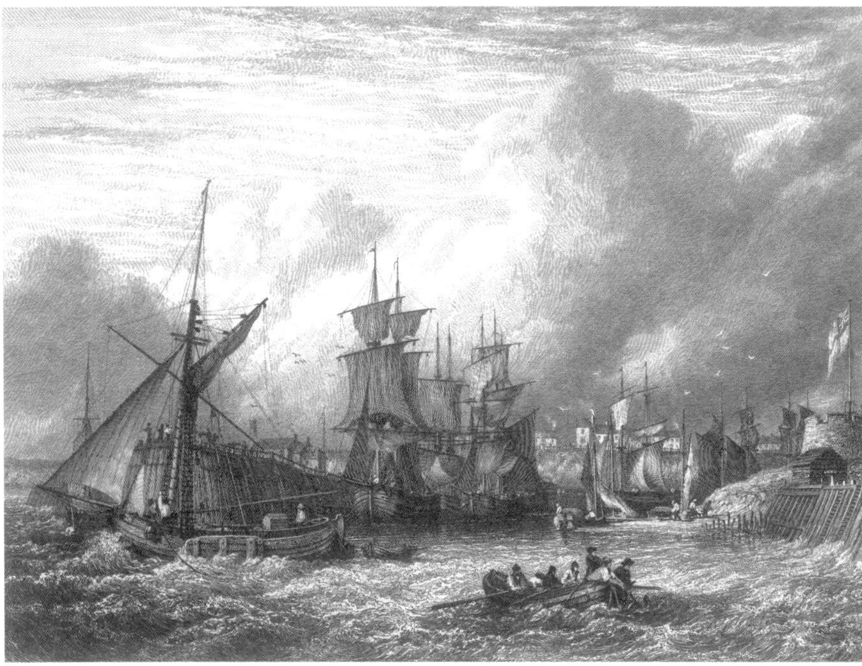

Fig. 3.2
James Stark, *The Mouth of the River [Yare]*, 1834, engraving by W. Miller, in J. Stark and J. W. Robberds, *Scenery of the Rivers of Norfolk* (Moon, Boys and Graves, 1834)

Ocean'.[4] This was a name that persisted until the conflict with Germany at the time of the First World War; it was the Dutch who generally referred to the sea as the 'Noordzee' or in English, the 'North Sea'. What Robberds regarded to be a more accurate representation of the extent of the estuary in Roman times was constructed by delineating the putative extent of the estuary on the basis of the extent of the marshes, meadows and valleys in the 1797 map of Norfolk by William Faden (1749–1836).

Robberds would, over the next few years, collaborate with the artist James Stark (1794–1859) on the first pioneering study of Broadland and the rivers and valleys of eastern Norfolk. This interdisciplinary study was published in parts between 1828 and 1834 and contains 36 engravings that depict the scenery of the former estuary as it was in the early nineteenth century.[5] In the text, Robberds introduces us to the estuary of Roman times, known once as 'Garruenos', while Stark depicts in an engraving, *The Mouth of the River [Yare]* at Great Yarmouth (1834), 'all that now remains of a once extensive estuary' (fig. 3.2). In this illustration of a stormy landscape at the mouth of the Yare, we see colliers waiting for a favourable wind, wherries being loaded with sea sand and the Yarmouth Beachmen, a volunteer rescue and salvage crew, perhaps on their way to give assistance to the single masted coastal vessel on the bar. It is a vivid scene.

James Stark was one of the artists brought together between 1803 and 1833 by the Norwich Society of Artists under the initial leadership of John Crome (1768–1821) and Robert Ladbrooke (1768–1842).[6] As a group, most of the artists, often now referred to as the Norwich School of Artists, were landscape painters and together they produced a substantial body of work inspired by the Norfolk landscape, whether it was the woodlands and heaths or rivers and coasts. Favourite haunts for them on the coast were Yarmouth and Cromer. There they painted seascapes, boats, jetties, quays, markets, activities on the beach and general coastal landscapes. A number of the pictures depict boats flying the Dutch flag, close to shore, symbolic of the close ties between Norfolk and the Low Countries in times of both peace and war. The Dutch transformed Yarmouth's herring fishing industry through the development of new fishing techniques and curing processes, and it was a Dutch engineer, Joas Johnson, who finally secured the entrance to the harbour in the late sixteenth century and prevented it from continually silting up.[7]

The influence of the Dutch, however, extended far beyond the coast. Immigrants from the Low Countries, 'Strangers', at one time made up about a third of the population of Norwich and had a particularly significant impact on the development of the textile industry.[8] More importantly though, in the context of this essay, was the influence of paintings by the seventeenth-century Dutch masters of landscape painting, such as Meindert Hobbema (1638–1709) and Jacob van Ruisdael (1628/9–1682; fig. 3.3). Their naturalistic style, together with the lowland landscapes they painted, clearly resonated with the Norwich artists.

The close links between Yarmouth and the Dutch are celebrated in the painting of *The Dutch Fair at Great Yarmouth* by George Vincent (1796–1832) from 1821 (fig. 3.4). The Dutch Fair was held on the Sunday before 21 September and marked the arrival of Dutch fishing vessels for the start of the herring season. In the painting, a number of boats can be seen beached on the shore under the gaze of Britannia atop the newly constructed (1817–19) memorial to Lord Nelson. The bustling scene depicted by Vincent might however be something of a picturesque contrivance as the description of a fair in 1785 has the Dutch boats moored along the South Quay.[9]

Great Yarmouth provided not only numerous subjects for the Norwich School of Artists, but also a home for some of them, including John Sell Cotman (1782–1842), one of its leading members. He also depicted a scene with the memorial to Nelson visible in the distance, but in this case the monument very much represented Britain's naval superiority at the time. Unlike Vincent, who

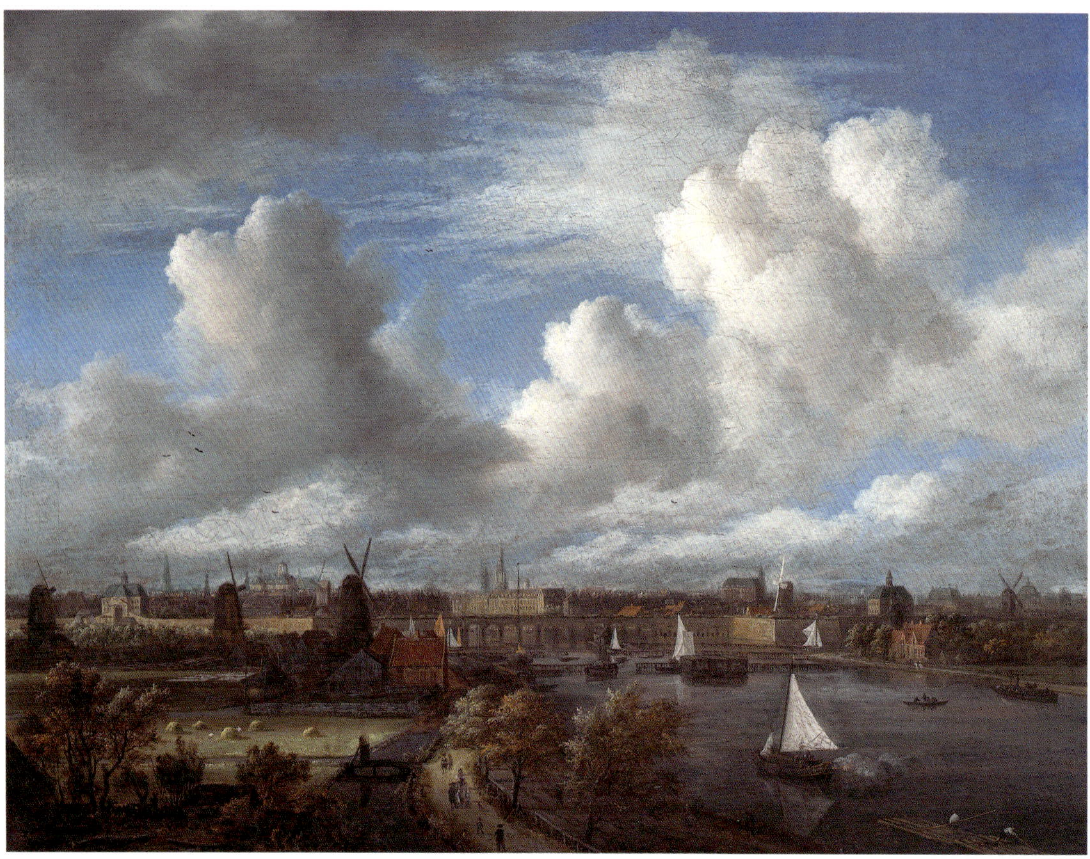

Fig. 3.3
Jacob Van Ruisdael, *Panoramic View on the Amstel looking towards Amsterdam*, oil on canvas, *c.*1675–80. The Fitzwilliam Museum, University of Cambridge

was celebrating the close ties between the British and the Dutch, Cotman's painting was of *Dutch Boats off Yarmouth, Prizes during the War* (*c.*1823–4). There had been a number of Anglo-Dutch wars during the seventeenth and eighteenth centuries and Cotman would have been aware of a number of conflicts during the Napoleonic era.

A much quieter scene of boats beached on the Norfolk coast is depicted by Cotman in his oil *Seashore with Boats*, painted in the first decade of the nineteenth century (fig. 3.5). The bold use of colour and its almost abstract design is reminiscent of some of his best watercolours from the same period. The painting vividly depicts, through the use of colour, how the sediments of the beach are related to and nourished by the sands and gravels of the cliffs, capturing their essence.

That essence derives from the cliffs of Norfolk being predominantly made up of till, a mixture of clay, sand, gravel and boulders, deposited during successive

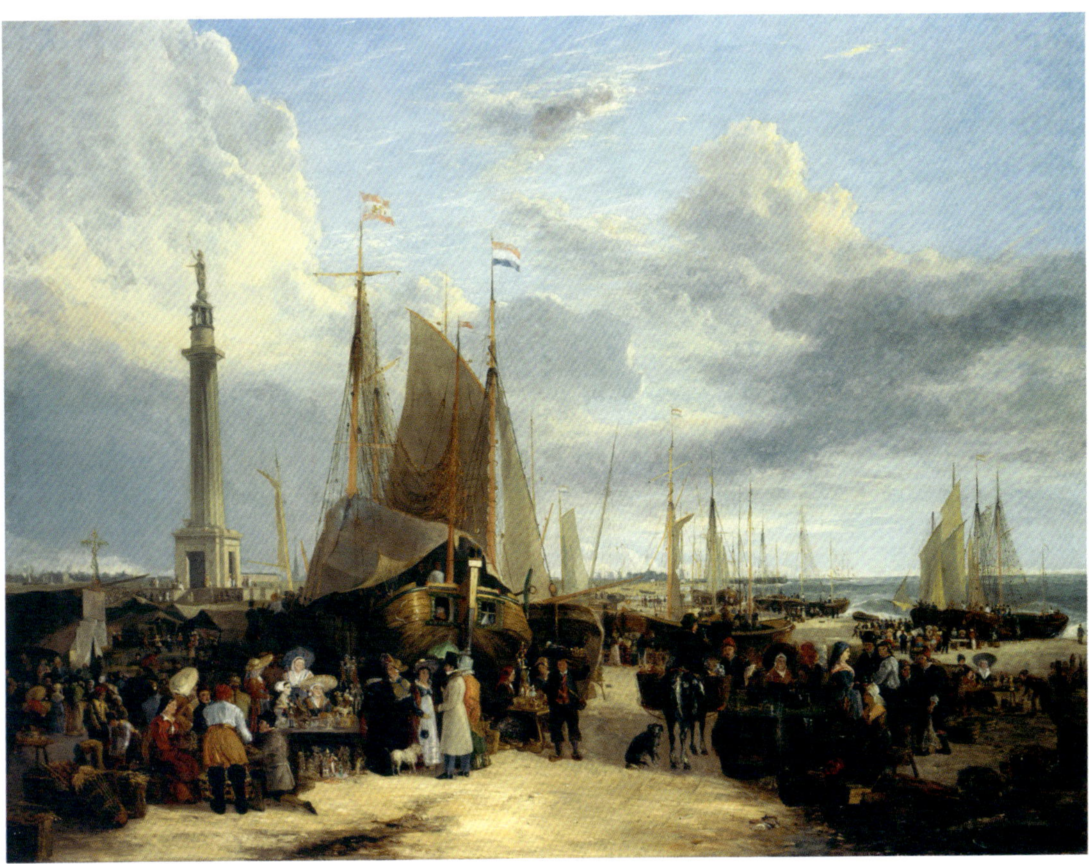

Fig. 3.4
George Vincent, *The Dutch Fair at Great Yarmouth*, 1821, oil on canvas. Time and Tide Museum of Great Yarmouth Life

glaciations. Indeed, some of the highest cliffs are an exposed part of the Cromer Ridge, a particularly notable terminal moraine, reaching heights of just over 100 metres and formed during the Anglian glaciation some 450,000 years ago. Robert Ladbrooke painted *Beeston Regis from the 'Roman Camp'* (date unknown) on the Cromer Ridge and depicts two circular hills, or kames, that are steep-sided mounds of sand and gravel deposited by the melting ice sheet. One of those has now disappeared and the other, known as Beeston Bump, is rapidly eroding; the summit is now only a few metres from the cliff edge. That erosion of the unconsolidated till at Beeston and elsewhere has resulted from both wave action at the foot of the cliff and slumpage at the top, typically associated with saturated soils and heavy rainfall.

The coastal scenes depicted by the Norwich School of Artists in the late eighteenth and early nineteenth centuries provide us with a picture of the coastal landscape before any sea defences were installed. Without defences, recession rates averaged about one metre a year, resulting not only in the loss of one of

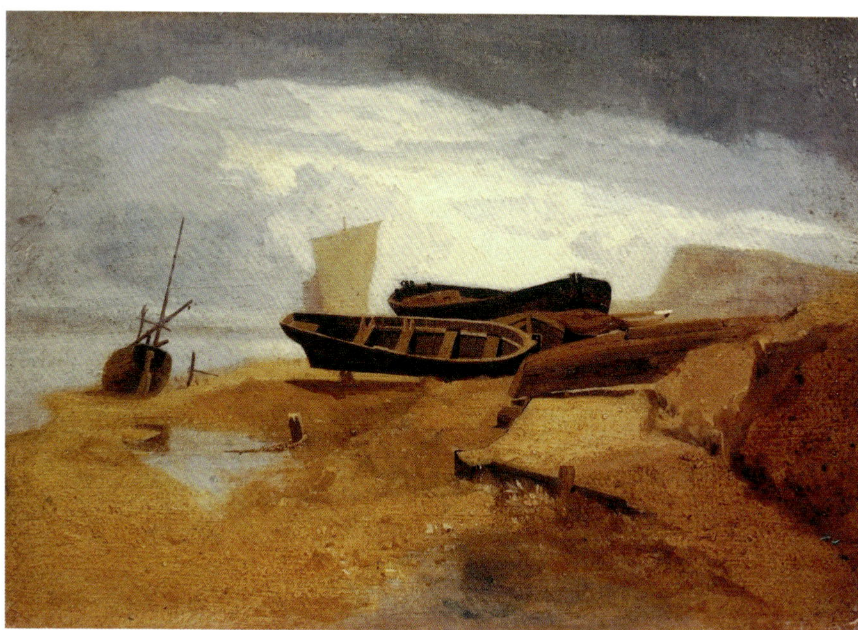

Fig. 3.5
John Sell Cotman,
Seashore with Boats,
1808, oil on board.
Tate, London,
purchased 1935

the Beeston hills and other stretches of the coast but also in the loss of a number of coastal settlements, such as Shipden and Eccles. It was not until the middle of the nineteenth century that they started to build defences, characterised by wooden groynes, sea walls and embankments.

Over time, vulnerable places were gradually defended and, following the devastating floods along the east coast of England in 1953, the government financed the building of hard defences to protect stretches of the cliffs and

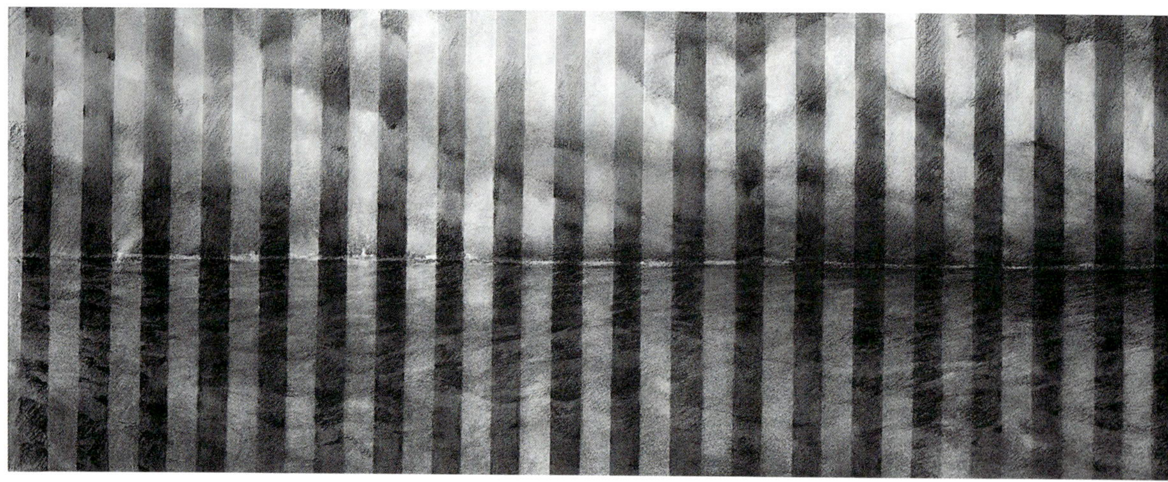

low-lying areas, committing it to long-term maintenance and giving planning authorities a false sense of security. Developments were consequently allowed to the edge of the coast, ignoring the natural processes that shape it. This was an attempt to make the line of the coast permanent, but it was not to last.

Coastal managers now face a wide range of problems including the accelerating rise in sea levels as a result of global warming, coastal squeeze between a rising sea level and hard defences, the high cost of maintaining sea defences and the question of how to support communities threatened by coastal erosion and flooding. The pressures exerted on the coast are powerfully captured in the charcoal image by Jayne Ivimey (b.1946), *Coastal Squeeze* (2024), which depicts the pressures on the narrow coastal zone by both the sea and sky from waves and rain (fig. 3.6). The vertical bars allow the viewer to capture a sense of movement in the sea and sky at day and night, but they can also be seen as representing the hardened coastline of seawalls, promenades, groynes, revetments and rock armour. These are seen as a modern-day 'corset', preventing the coast from breathing and responding to natural processes.

Current shoreline management plans now recognise that we cannot hold the line along the entire coast; the environmental and financial costs are too high. Certainly, there is a need to defend major settlements and infrastructure where we can, but there is also a need to return sections of the coast to a more natural state.[10] Inevitably properties will be lost to the sea. Julian Perry (b.1960) captures the moment of trauma when property is lost during a violent storm in *Fanfare 34* (2010), a quiet, still and somewhat surreal image of a static caravan floating above the sea, detached from the cliff where it might once have stood (fig. 3.7).

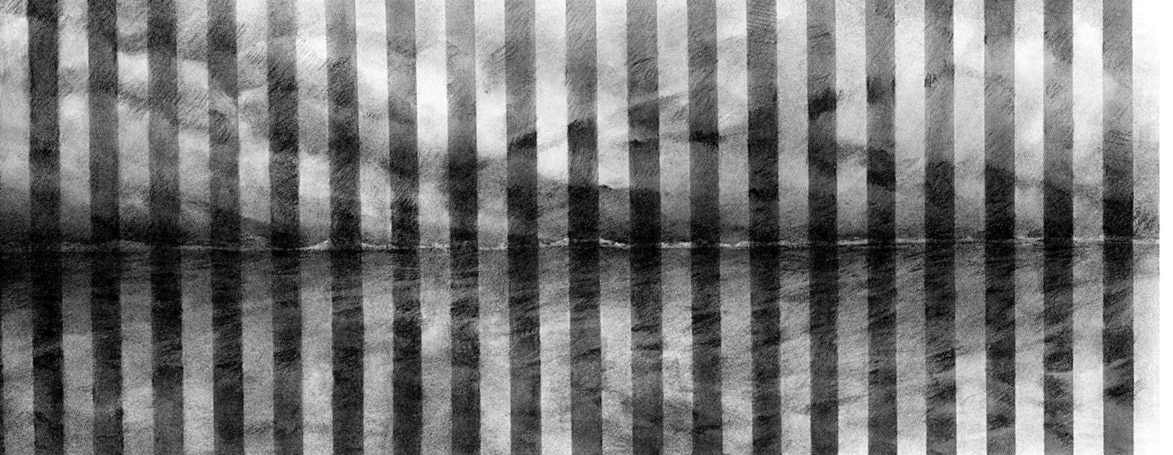

Fig. 3.6
Jayne Ivimey,
Coastal Squeeze, 2024,
charcoal on paper

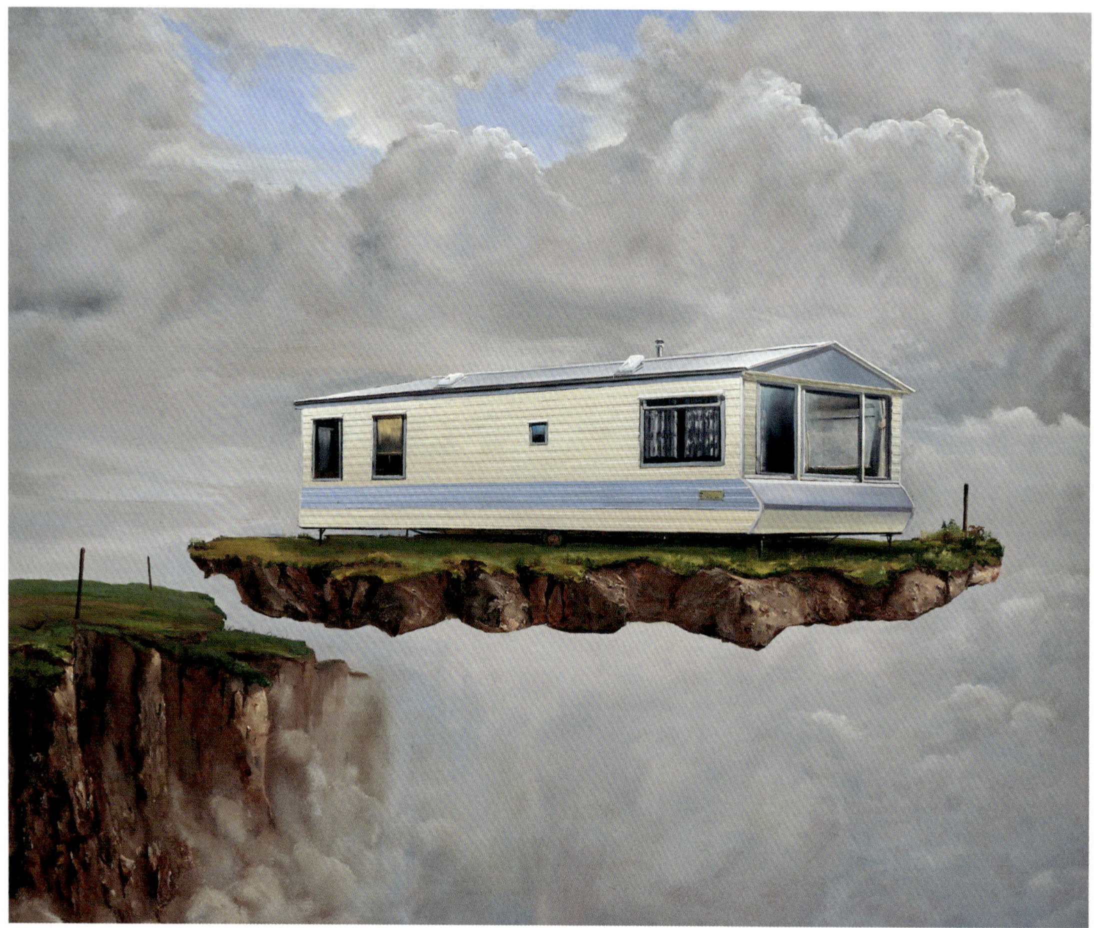

Fig. 3.7
Julian Perry,
Fanfare 34, 2010,
oil on panel

The line of the coast is being redrawn again and will continue to change as thermal expansion of the oceans and melting of ice sheets and glaciers result in sea levels potentially increasing by as much as a metre by the end of the century. Our remote ancestors experienced such change with the flooding of Doggerland, but modern humans have not. It is in our hands to mitigate these changes by reducing carbon emissions and by adopting adaptation strategies to increase our resilience to future sea level rise. This is within our grasp but requires a collective commitment.

‘THE LEAVING OF KATTEGAT SEA (OR THE MERMEN’S LAMENT)’

HARUN MORRISON

Notes for 'The Leaving of Kattegat Sea'

These lyrics were written with the singers of the Slupkoret Ebeltoft (a Danish sailing club choir; fig. 4.1) in mind and overwrite those of the folk song 'The Leaving of Liverpool'. Rather than an ensemble of retired yacht owners, sailors and naval men, could the lyrics imaginatively reframe the singers as mermen? Could the singing of the song shift their position and relationship to the sea?

The Kattegat Sea has been the focus of environmental debate, particularly in relation to contaminated soil linked to a company that has conveniently declared itself bankrupt: Nordic Waste.[1] Replacing the anthropocentric and capitalist perspective that causes such destruction with the outlook of the merman, I reference a 400-year-old Greenland shark, which the visual artist Marie Kølbæk Iversen (b.1981) notes is called either 'merman' or 'mermaid' across the North Germanic languages.[2] Its traditional Danish name is '*havkal*', that is 'merman'. This 'merman-shark' is older than the Treaties of Westphalia (1648) responsible for formalising the notion of the nation state. I see the Greenland shark not as floating, but as submerged and circulating; a signifier of the navigation of the seas prior to artificial national boundaries and maritime laws for trade.

Fig. 4.1
The Slupkoret Ebeltoft sailing club choir, 2024

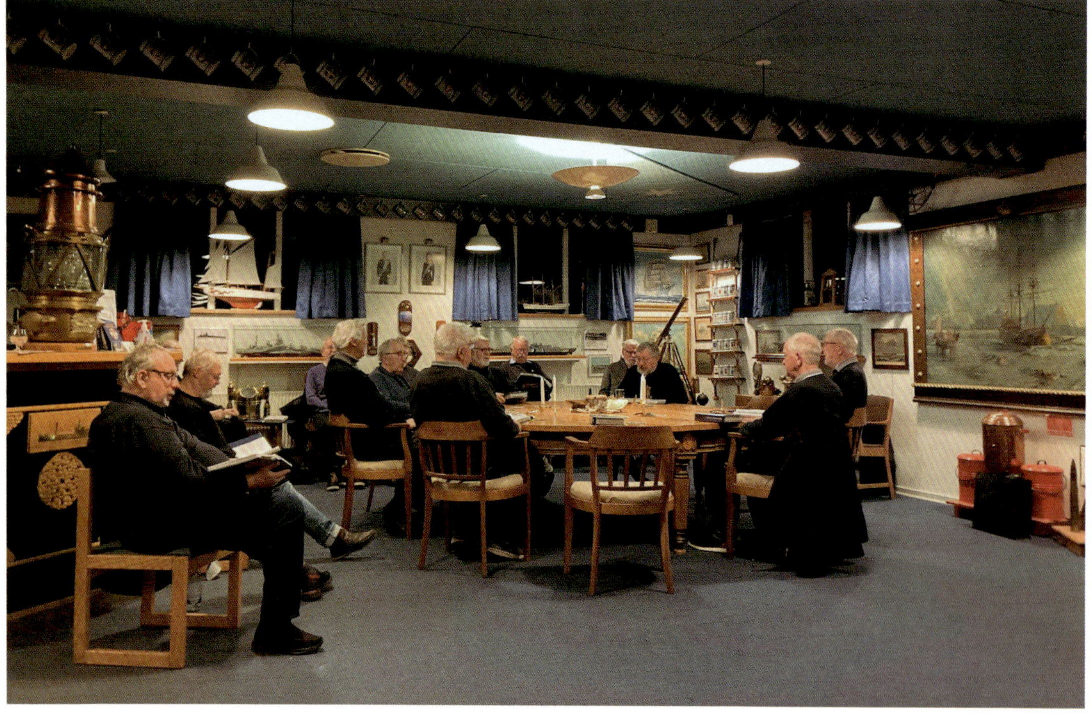

'THE LEAVING OF KATTEGAT SEA (OR THE MERMEN'S LAMENT)'

Melody: 'The Leaving of Liverpool'
Lyrics commissioned by Hosting Lands, 2024

[1] Goodbye to modern waters
Kattegat Sea, you were home.
It's time to explore the land.
Walk on stones, not swim in foam.

Refr. We Mermen love the Kattegat
Though we leave we hold the sea inside
Each tear finds the Baltic, keeps rolling to the shore
Wet eyes on cobbled streets of Ebeltoft

[2] Goodbye seagrass, wave us back
Shipwrecks we made our shelter
Relentlessly you dredge our seabed
Who ashore will give us answers?

Refr. We Mermen love the Kattegat ...

[3] We're bound for the Capital
We'll sing to Copenhagen
Campaign in pain for the sea we love
No home till we're heard

Refr. We Mermen love the Kattegat ...

[4] Where have the grey pup seals disappeared?[3]
Past centuries they were plenty
We mourn quadricentennial sharks
Older than the Westphalia Treaty

Refr. We Mermen love the Kattegat ...

[5] Every minute away from sea
We lose strength we need your promise
If we cannot make it to Copenhagen
We ask of you – sing for us.

Refr. We Mermen love the Kattegat ...

'THE IMAGE OF ETERNITY': DEEP TIME, POWER AND VULNERABILITY IN REPRESENTATIONS OF SEAS AND OCEANS

COURTNEY TRAUB

Thou glorious mirror, where the Almighty's form
Glasses itself in tempests; in all time
Calm or convulsed-in breeze, or gale, or storm,
Icing the pole, or in the torrid clime
Dark-heaving; boundless, endless and sublime-
The image of eternity ...

Lord Byron, 'Childe Harold's Pilgrimage' (1812–18)[1]

The sea is history.

Derek Walcott, 1978[2]

When did oceans and seas become such potent symbols of eternity – of forces stronger and far more dangerous than any technological power humans might harness in an effort to influence them; of time scales that stretch so far beyond our own that we strain to even imagine them? In Western literary and artistic traditions the answer might be 'at least since Antiquity, if not earlier'. In Homer's *Odyssey* and the biblical tale of Noah and the great flood, the ocean is a potent protagonist and foe; a site of terror, spiritual conquest and awe-worthy power. But it was the Romantic period in art, literature and philosophy – stretching roughly from the Revolutionary 1790s in Europe to around the 1830s – that generated some of our most enduring ideas about watery bodies (and how we relate to them). Philosophers and writers including Immanuel Kant (1724–1804), Edmund Burke (1729–1797), Samuel Taylor Coleridge (1772–1834) and Lord Byron (1788–1824), as well as European painters such as Caspar David Friedrich (1774–1840), J. M. W. Turner (1775–1851) and Théodore Géricault (1791–1824), pondered the disquieting power of seas and oceans through new aesthetic and poetic vocabularies. These explored how human encounters with such powerful, and ostensibly eternal, natural entities arouse feelings of profound vulnerability and mortal terror – ones that many Romantics would relate to an experience of 'the sublime'.

In simplified terms, the sublime might be understood as an aesthetic approach, or affective response, that grapples with feelings of awe, overwhelm and/or terror in the face of something so large or complex that it can't be understood or represented as a whole *thing*. Edmund Burke, the English essayist and anti-revolutionary, put it this way:

> *The passion caused by the great and sublime in nature ... is astonishment: and astonishment is that state of the soul in which all its motions are suspended, with some degree of horror.*[3]

For Burke, rugged and enormous natural objects, including oceans, as well as extreme weather conditions like storms, are among the phenomena that spur instinctive feelings of astonishment and terror in human minds.[4]

Enter the age of the Anthropocene, a geological era triggered by human activity, which many scientists argue began around 1950.[5] In the face of global climate change and other environmental threats, recent cultural and artistic responses to waters that are transforming more rapidly than we can wholly register – bodies that encroach, menace and shapeshift even as we vitally depend on them for our survival – are consciously and unconsciously tied to earlier Romantic notions of sublimity. From popular films and TV series such as *The Day After Tomorrow* (2004) and *Foundation* (2021) to photography and literature, narratives about menacing, mesmerising seas and oceans have become central tropes of environmental anxiety. In particular, contemporary works of art that attempt to grapple with phenomena such as oil spills; with waters and ocean floors permeated with microplastics; or with the threatened swallowing of island nations like Tuvalu and Kiribati as sea levels rise, reinterpret sublimity in an age when 'nature' has never seemed more, well, *unnatural*.

As such, you might say that Romanticism has never really receded when it comes to how we tend to envisage, and *feel through*, our relationships to incommensurably powerful forces – including seas and oceans.[6] Timothy Morton uses the term 'hyperobject' to refer to objects and experiences that overwhelm our capacity for description and cognition; 'things that are huge and, as they say, "distributed" in time and space – that take place over many decades or centuries (or indeed millennia) and that happen all over Earth – like global warming. Such things are impossible to point to directly all at once.'[7]

Oceans, seas and their ecosystems – vastly complex, globally intersecting and famously hard-to-predict things – can be described as hyperobjects. But before delving further into contemporary narratives about watery bodies and how they reframe the sublime, let's first consider a few examples of Romantic-era literature and art – or what I'll call key 'texts' for how we imagine relating to powerful and dangerous waters.

Among these, one of the most influential and culturally enduring is Samuel Taylor Coleridge's lyric poem, 'The Rime of the Ancient Mariner' (1798), featured in the 2025 Sainsbury Centre exhibition *A World of Water* through an 1885 copy of the illustrated volume from the Royal Scottish Academy by David Scott (1806–1849). The poem's titular mariner and poetic narrator crashes a wedding to recount his perilous encounters at sea, describing life-threatening experiences and the death of some of his shipmates – not least through the poem's most familiar lines: 'Water, water everywhere / nor any drop to drink.'[8] While many have focused on supernatural events and Christian allegories in the poem, French artist Gustave Doré (1832–1883), in his 1877 woodcuts, makes the ocean itself the central object of the mariner's (and in turn the reader's) *frisson* of fear.[9]

The poem and the illustrations both imagine the sea, and particularly its stormy and icy states, as monstrously animated. 'The ice was here, the ice was there / The ice was all around: / It cracked and growled, and roared and howled, / Like noises in a swound!', recounts the Mariner, and with this image we're plunged into a landscape that's equally menacing and mesmerising: one whose uncanny noises suggest the inherent instability of ice.[10] This is perhaps part of what makes seas and oceans uniquely frightening: their chameleonic ability to change from calm to stormy, liquid to solid and back again.

One of Doré's woodcuts illustrating the encounter with Arctic ice in 'Rime' depicts a ship travelling through a narrow channel surrounded by what appear to be claw-like columns of icebergs; it invites comparison to German Romantic artist Caspar David Friedrich's painting of 1823–4, *Das Eismeer* (*The Sea of Ice*) (fig. 5.1). In Friedrich's own sublime Arctic, the eye first settles on what seems to be monotonous blocks of ice, before identifying fragments of a wrecked ship, entrapped and jutting out among glass-shard sharp icebergs. Here, the human-made object is seemingly engulfed and imprisoned by the natural.

Meanwhile, J. M. W. Turner made shipwrecks and dramatic storms central subjects, partly shaping broader European Romantic aesthetic conventions. His 1842 painting *Snow Storm – Steam-Boat off a Harbour's Mouth* (1842) exemplifies how he 'wanted us to feel the sea's power and volatility as if we were there, swept up in a swirling vortex of wind and waves or the foreboding bleak, open water' (fig. 5.2).[11]

In much of his later work, Turner drew the tightly symbiotic, but also often violent, relationship between humans and seas. In a painting entitled *Yarmouth Sands* (*c.*1840), one of several depicting maritime Norfolk landscapes, a ship is

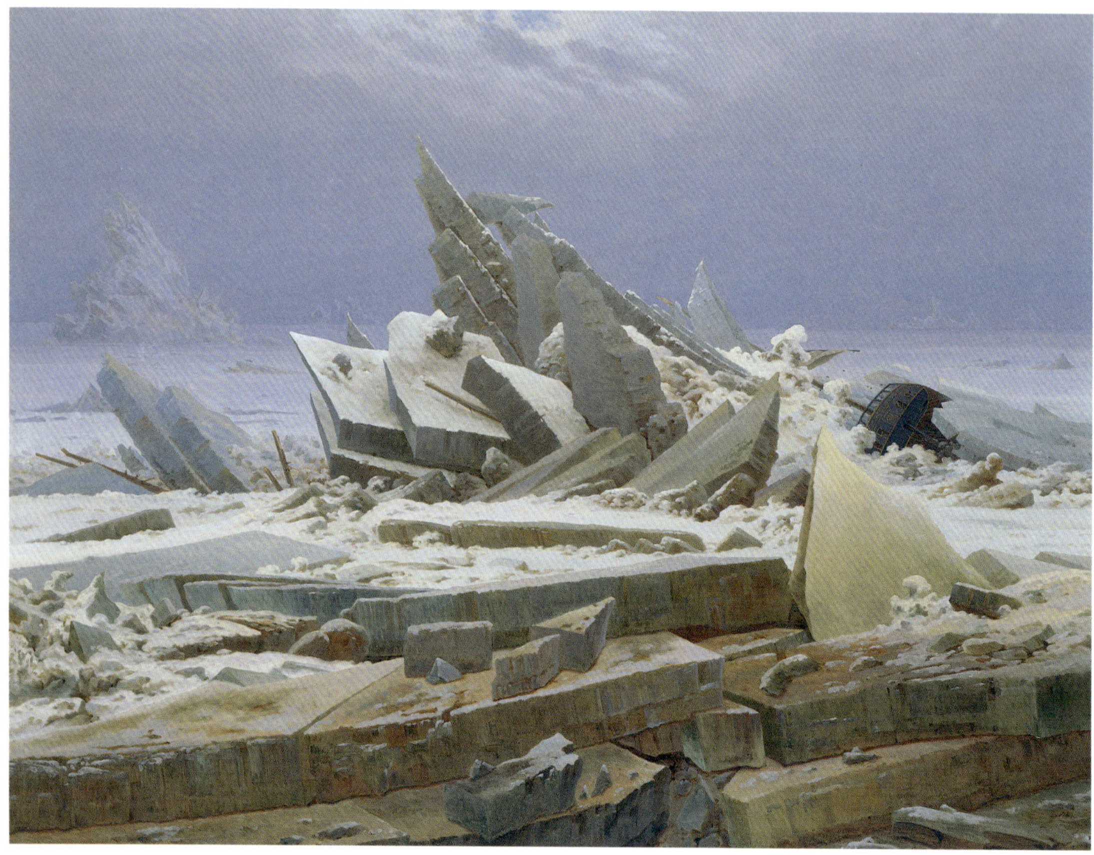

Fig. 5.1
Caspar David Friedrich, *Das Eismeer* (*The Sea of Ice*), 1823–4, oil on canvas. Hamburger Kunsthalle, Hamburg

dramatically tossed on the waves – representing a meditation, in part, on 'the frailty of human culture pitted against the forces of nature'.[12]

Meanwhile, an engraving from William Miller (1796–1882) entitled *Great Yarmouth, Norfolk, 1829 After J. M. W. Turner RA (1775–1851)*, produced *c.*1830 and shown in the exhibition *A World of Water*, depicts a stormy, windswept maritime landscape that visually blurs the lines between sea and land. Dark clouds loom overhead as small human figures and dogs comb the beaches and shallow waters; a fishmonger appears to survey her catch in the foreground. But despite the clear power and threat posed by the sea, human-built structures – a Nelson column, sea barrages and other buildings designed to withstand the changing tides, ships gliding through inlets – seem to triumphantly dominate. In addition to suggesting a symbiotic overlapping between sea and 'civilisation', this piece also intimates how human technology can overcome threatening natural forces, and even profitably use them. It thus envisages a tense power struggle between ferocious, timeless seas and the human will to conquer them – in line with

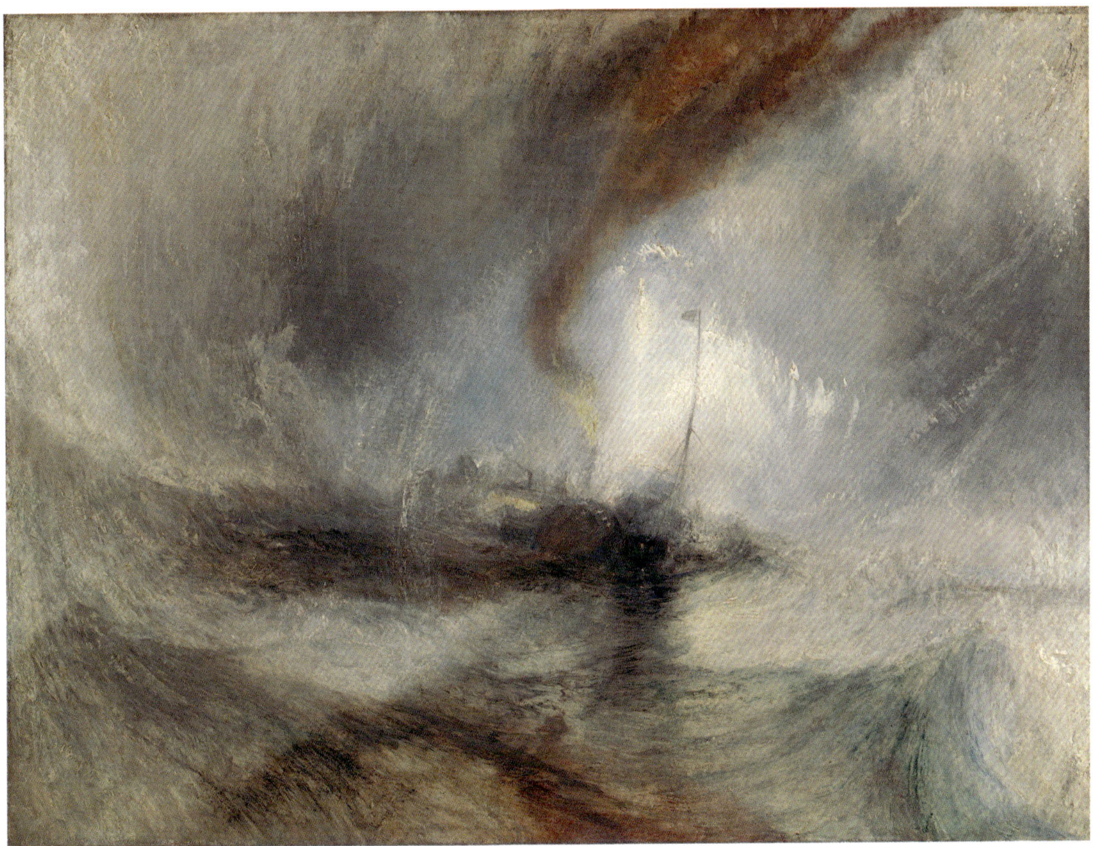

Fig. 5.2
J. M. W. Turner, *Snow Storm – Steam-Boat off a Harbour's Mouth*, 1842, oil on canvas. Tate, London, accepted by the nation as part of the Turner Bequest 1856

reigning imperialist and mercantilist constructions of 'nature' as something to be tamed. Entirely absent in such conceptions, of course, is a sense of how 'nature' itself might be profoundly altered by human activity.

CAN 'ETERNAL' SEAS SURVIVE US?

Indeed, one of the glaring blind points in many Romantic depictions of oceans and seas – including in the narrative poem from Byron that opens this chapter – is a notion that they are eternal and unchanging – and thus, by corollary, invulnerable to human activity. When Byron calls the oceans '[d]ark-heaving; boundless, endless and sublime / The image of eternity'[13] he is drawing attention, as Derek Walcott (1930–2017) does in 'The Sea is History', to the 'deep time' scales of oceans and seas. But Walcott's poem, notably by unearthing the traumas of the transatlantic slave trade, centres what Byron occludes: the sea is a 'grey vault' of human memory and activity, including as a route for the enslavement of millions.[14] While Walcott's poem can't be said to address histories of marine industrial pollution and ecological degradation, many would argue that the violence of slavery was inseparable from the environmental violence introduced by global mercantilism and, ultimately, consumer capitalism. In this sense, seemingly unrelated histories of global violence concomitantly mark, or even scar, seas and oceans – whether figuratively or literally.

But most Romantic-era writers and artists failed to bring to their renderings of sublime oceans and seas the fact that these watery bodies had already changed immensely under human influence – especially since the Industrial Revolution, but also prior to it. From early Dutch geo-engineering projects that drained wetlands and fens and reshaped coastal areas, to the current-day acidification and 'plastification' of oceans under the influence of climate change and consumerism, these bodies bear profound, enduring marks of human history. Their future, too, remains uncertain, with the risk of broad ecosystemic collapse in seas and oceans a real possibility as waters warm and acidify.

EARLY OBSERVERS AND CONTEMPORARY RESPONSES

Lest you think the aforementioned Romantics should be let off the hook for failing to see how humans were shaping 'nature', there were in fact a few early observers of such impacts; you might even call them whistle-blowers. One was the American essayist and Transcendentalist Henry David Thoreau (1817–1862). In works such as *The Maine Woods* (1864), he charted (with growing alarm) how the logging industry was thinning out once-thick forests on the Eastern seaboard and noticed the startling absence of formerly abundant species in local rivers and marshes: something we would call 'biodiversity loss'. In *Man and Nature* (1864), the naturalist and geographer George Perkins Marsh (1801–1882) painstakingly catalogued how human activity was already reshaping the planet, including seas and oceans. Marsh notably documented the 'interference of man with the order of nature' in the Zuiderzee, a portion of the now-Netherlands that was initially transformed from 'a marsh into an open bay' sometime in the fifth century; interestingly, he advocated for it to be drained so that the 'coast currents be restored approximately to the lines they followed fourteen or fifteen centuries ago'.[15] Marsh's accounts of how humans modify coastlines, and their attendant ecosystemic effects, were remarkably prescient.

These early observers laid some of the groundwork for twentieth-century conceptions of ecosystem balance and disruption, anthropogenic changes to Earth systems and other central ideas in contemporary ecology and philosophy. The term 'deep time', widely attributed to the eighteenth-century geologist James Hutton (1727–1797), is now in regular usage; it suggests just how far back we have to imaginatively stretch to fathom the speed with which we have managed to change Earth systems over the past few decades – and how far into the future something like global heating will continue to impact the planet.[16]

These concepts also abound in contemporary narratives and artistic works exploring our relationship to oceans and seas. The advent of the field known as Blue Humanities 'invites exploration of cultural connections between people and oceans, past and present'.[17] This is also the focus of much recent work from artists with interests in explicitly addressing the environmental health of seas and oceans.

US-based artist and activist Bonnie Monteleone's travelling exhibition, *What Goes Around, Comes Around*, is one intriguing example among many others. Monteleone's large-scale canvases reflect on the permeation of plastics and microplastics in contemporary oceans and seas, adapting the iconic nineteenth-century woodblock print by Japanese artist Hokusai (1760–1849), *Under the Wave of Kanagawa* (1831), to pose powerful questions around just how much human activity has changed the oceans – perhaps irreversibly (fig. 5.3). Her series supplants, in a sense, the 'traditional' sublime of the original, which shows an enormous wave threatening to engulf three boats and their occupants. In Monteleone's renderings of Hokusai's wave, it is, instead, a 'toxic sublime' that dominates our awareness; vividly hued plastic assemblages bending into wave shapes draw us in, but also invite uneasy reflection. As one reviewer put it, surveying the tableaux, which together suggest waters progressively saturated by plastic particles:

> *The viewer gets a sense of how much our oceans have changed since Hokusai's vision of it, nearly 200 years ago. Next to each canvas is a bin with some of the objects depicted in the artwork that visitors can pick up and tangibly interact with.*[18]

Plastics take on a monstrous sort of *agency* in this series; they become, to borrow from Jane Bennett, a sort of 'vibrant matter' that – though produced by us – has in turn come to penetrate, colonise and transform our world, including seas, oceans and even our bodies.[19]

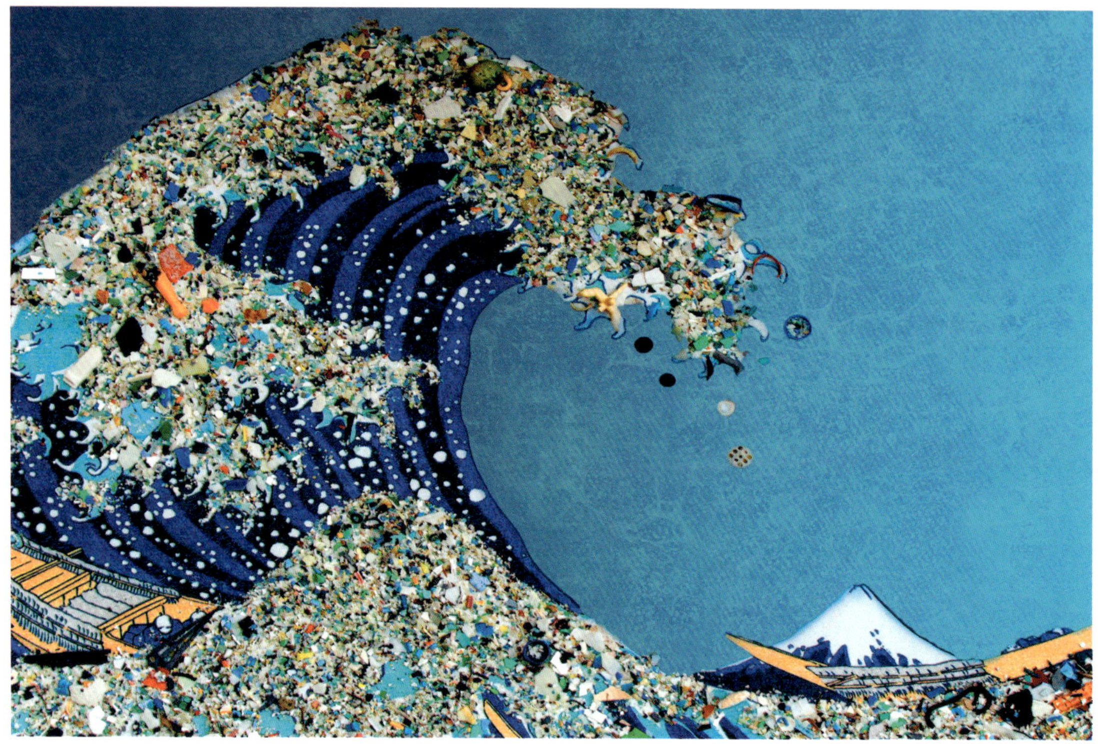

Fig. 5.3
Bonnie Monteleone, *Plastic Ocean: In Honor of Captain Charles Moore*, 2011, mixed media

Monteleone is not the only contemporary artist to draw on Hokusai's iconic 'Wave' in a bid to draw attention to toxified oceans and seas. As Stefan Helmreich notes:

> *[A]daptations of Hokusai's 'Wave' these days ... increasingly point to more general anxieties about catastrophic climate change and to worries about ocean pollution, acidification, and plastification. In such usages, the 'Wave' operates as a synecdoche for, a symbolic capture of, the difficult-to-apprehend vastness of the ever-moving, interconnecting, and possibly threatening sea.*[20]

Meanwhile, deep-time scales intersecting with the runaway melting of centuries-old icebergs are the subject of French-Swiss artist Julian Charrière's project *The Blue Fossil Entropic Stories,* featured in the *A World of Water* exhibition. In 2013, Charrière (b.1987) travelled to Iceland, where he climbed an Arctic iceberg and applied a gas torch to melt it below his feet for eight hours. The project saw Charrière confront 'the elements in a seemingly hopeless battle – human time against geological time'.[21]

The artist, who frames the project as 'a kind of contemporary version of Caspar David Friedrich's *Wanderer Overlooking the Sea of Fog* ... and a questioning of our relation to nature as inherited from the Romantics via ecological thought', documented the results of his experimental performance in a series of three key images (fig. 5.4).[22] These activate both the stark, traditionally sublime beauty of icebergs framed by a twilight-bathed sea, and a more recent form of cognitive overwhelm: how to quantify, measure or predict what icebergs might do next under the influence of human-induced warming? How might these 'hyperobjects' radically reshape coastlines and beyond as they melt – and how quickly? How much agency do we really have to influence the course we may have locked in for them by continuing to burn fossil fuels? All these questions are invited by Charrière's project.

The work of Canadian photographer Edward Burtynsky (b.1955) offers a final example of how contemporary artists reference, and critique, the Romantic sublime in representations of ecological catastrophes involving seas and oceans. Images of landscapes such as the Gulf of Mexico, where an oil slick in 2010 was made strangely and disquietingly beautiful when viewed from above, illuminate Burtynsky's approach to capturing the 'toxic sublimity' of ocean pollution as part of his multidisciplinary joint project on the Anthropocene (fig. 5.5).[23]

Fig. 5.4
Julian Charrière,
The Blue Fossil Entropic Stories III, 2013,
C-print

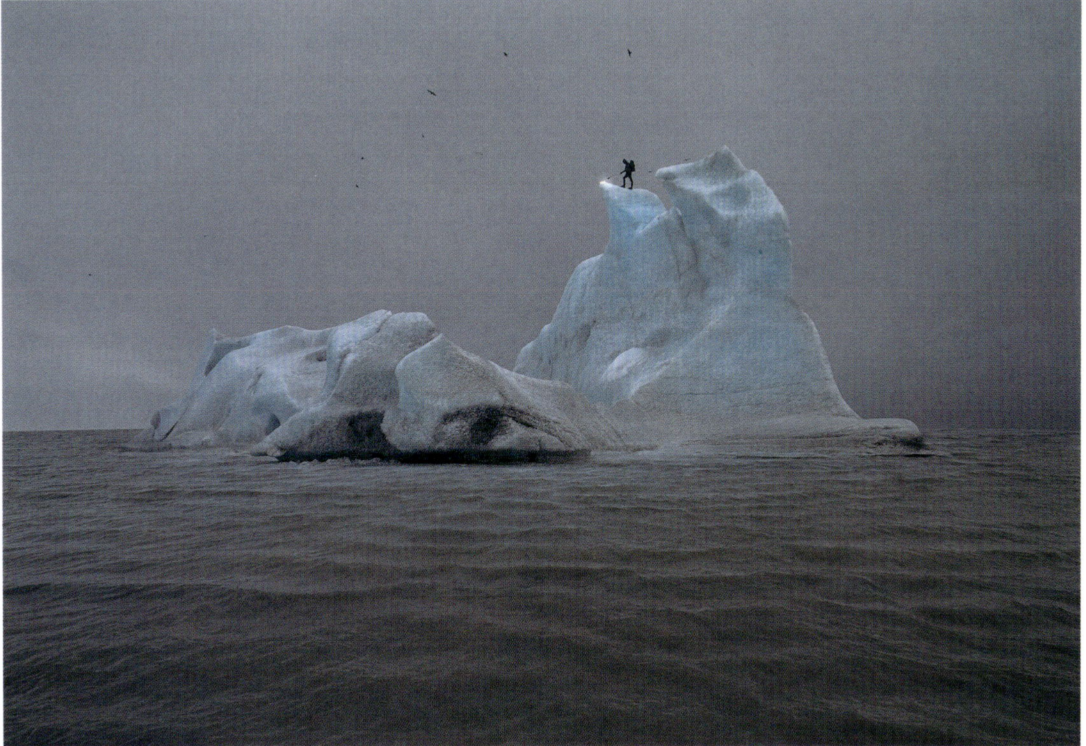

In the aerial image, a black oil slick bends and curves through an otherwise deep-green Gulf, creating forms that resemble earthquake fault lines, or deep scars on a body. The composition is hard to look away from, and that's precisely the point. What notably strikes us as awe-inspiring, and maybe even uncanny, is the hard-to-separate-out natural and 'unnatural' beauty of the landscape. But in the Anthropocene, using these terms in traditional and binary ways seems increasingly problematic anyway. After all, if a bee uses remnants of plastic to help build its hive, should we now characterise the hive as 'unnatural'? Similarly, if ocean ecosystems dramatically change and evolve in response to climate change and other human impacts, becoming more acidic, warmer and favouring fewer species, are they then forever tainted, or even monstrous? These are questions that can no longer be ignored.

What all these contemporary responses have in common is that they play with (Romantic) conceptions of sublime, watery bodies and the sense of deep

Fig. 5.5
Edward Burtynsky, *Oil Spill #10, Oil Slick, Gulf of Mexico*, 2010, C-print

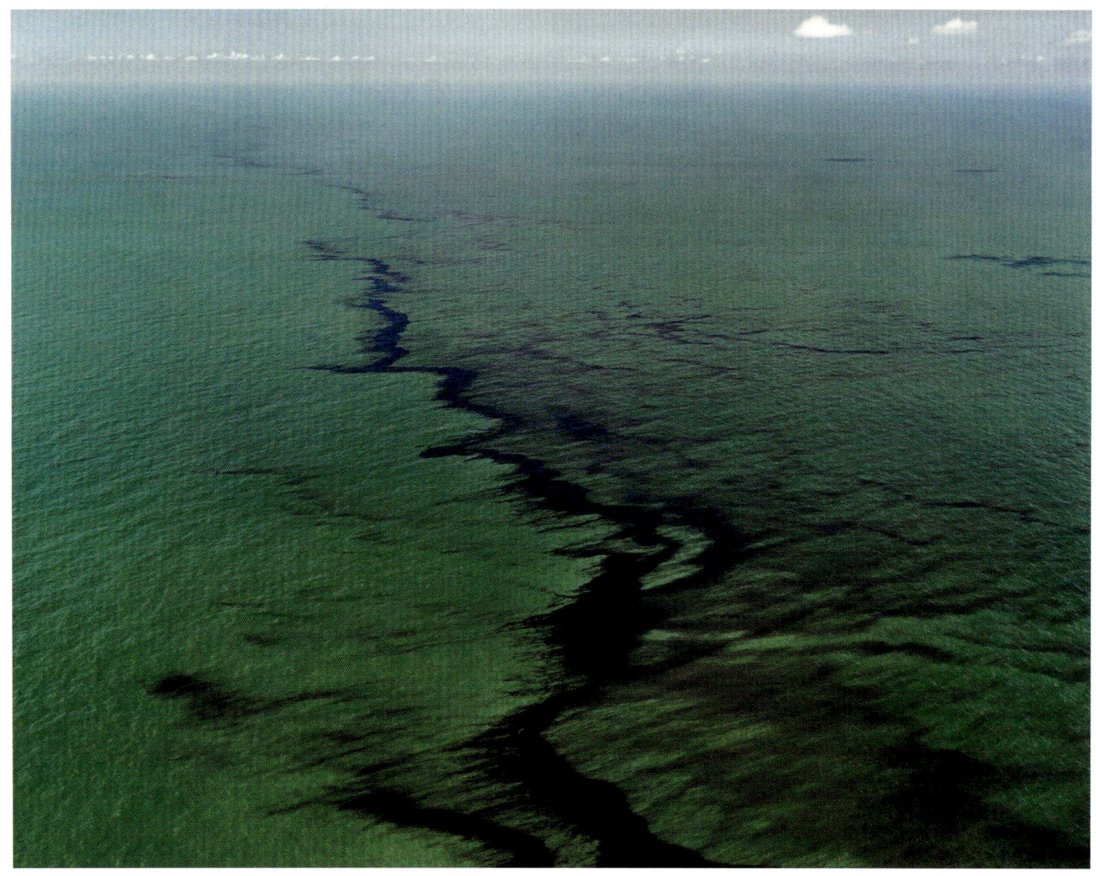

vulnerability they inspire in us, while raising startling new questions. These works, and many others, provocatively question these complex legacies, offering audiences arresting and immediate entry points for contemplating how we might best approach the ecological challenges that currently confront us.

By looking back at how Romantic art and literature framed seas and oceans as objects of awe-inspiring power, and then casting a critical eye on some of the ecological blind spots in those accounts, we might foster a deeper understanding around how we relate to these vital Earth systems in the present day – and how to avoid the pitfall of assuming them eternal and invulnerable.

TIPPING POINTS: PAUSING THE UNSTOPPABLE IN THE WORK OF MAGGI HAMBLING, CLAIRE CANSICK AND MARGARET MELLIS

ANTONIA BLOCKER

The urge to slow the passage of time is widely recognisable. As our lives tumble by, most can relate to the desire to pause certain moments and find a few more hours in the day. As the climate emergency unfolds, this desire takes on a new significance. No longer a simple, individual wish, but a universal need to give humanity more time to get things right, lest we barrel past a point of no return. According to the Intergovernmental Panel on Climate Change (IPCC), tipping points are 'critical thresholds in a system that, when exceeded, can lead to a significant change in the state of the system, often with an understanding that the change is irreversible'.[1] A number of different 'tipping elements' were identified at the turn of the century, which broadly fell into the categories of cryosphere, ocean-atmosphere and biosphere. Since then, thanks to scientific and technological advances, our understanding of these processes is more nuanced, and more frightening. The melting of ice sheets, permafrost loss and the dying of rainforests and coral reefs indicate a 'tipping cascade', the result of which would be a domino effect, the impact of which we can only begin to imagine.

The works I explore in this text share a common ambition, to capture an unstoppable moment and pause the uncontainable within the parameters of an artwork. From the explosive *Wall of Water* series (begun in 2010) by Maggi Hambling (b.1945) to the close and careful study of the sea of her native Norfolk by Claire Cansick (b.1971) and the abstract driftwood assemblages by Margaret Mellis (1914–2009), each artist attempts to arrest nature's flow. I appreciate that this desire may seem unrelated to our individual and collective relationship to, and involvement in, these environmental 'tipping points'. The cacophony of a smashing, splintering wave or the surge of a North Sea swell rendered in paint may seem insubstantial in the face of such a crisis. Nevertheless, perhaps these paintings belie a greater power, their potential ripple effect enough to affect awareness and translate a wish for more time into a behavioural shift.

Maggi Hambling's *Wall of Water* paintings are an enigmatic series, each canvas slightly larger than human scale, measuring some two metres in height (fig. 6.1). Suggesting spectral figures or fictional cities from afar, their subtle low horizon lines and white backgrounds create an immediacy of perspective. The dynamic fury of their colour and form at once gorgeous and confronting, each piece threatens to engulf the viewer in a tide of inky blacks, rich blues and startling pinks and ochres.

Hambling started the series after witnessing giant storm waves pounding the sea wall in Southwold, Suffolk. Comparing the North Sea to a 'raging beast', Hambling

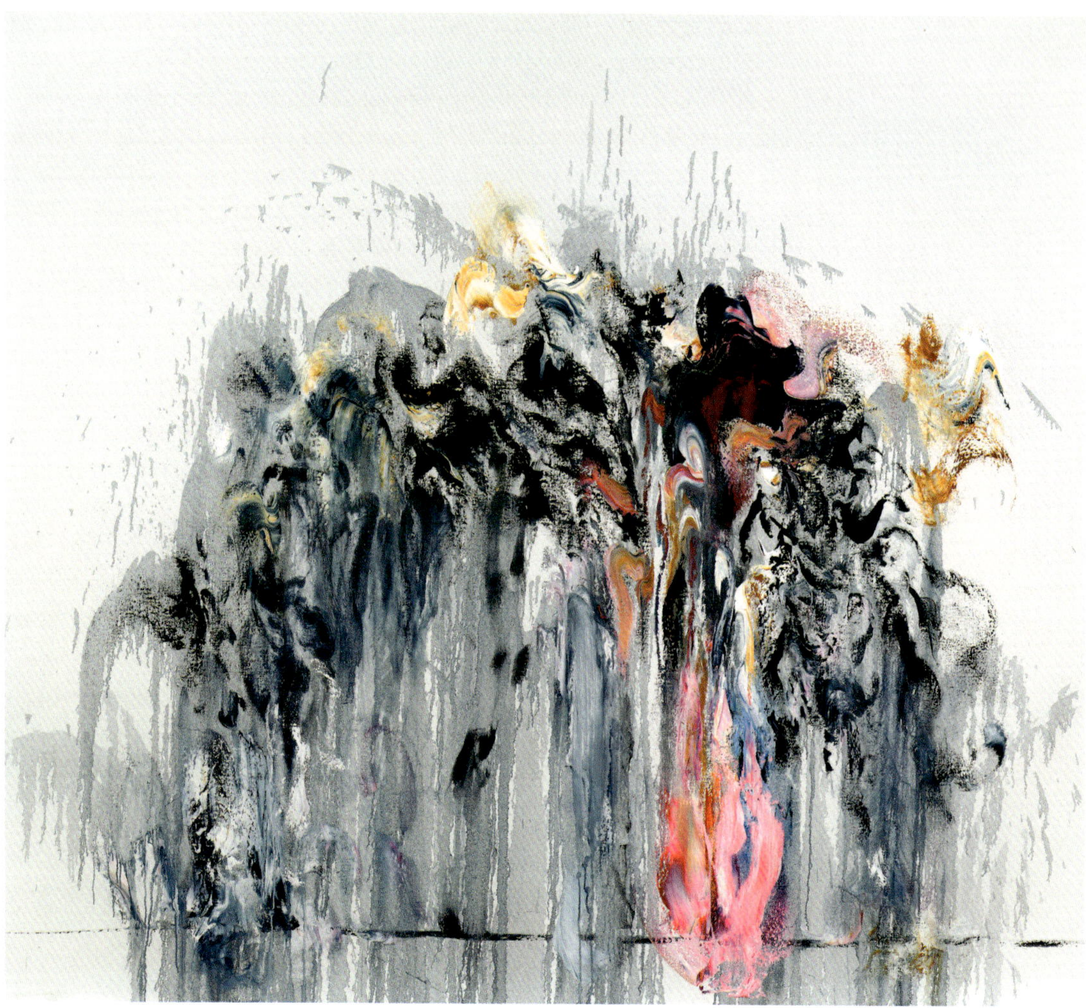

Fig. 6.1
Maggi Hambling,
Wall of Water VIII,
oil on canvas, 2011

Fig. 6.2
Maggi Hambling,
Erosion, oil on canvas,
2022

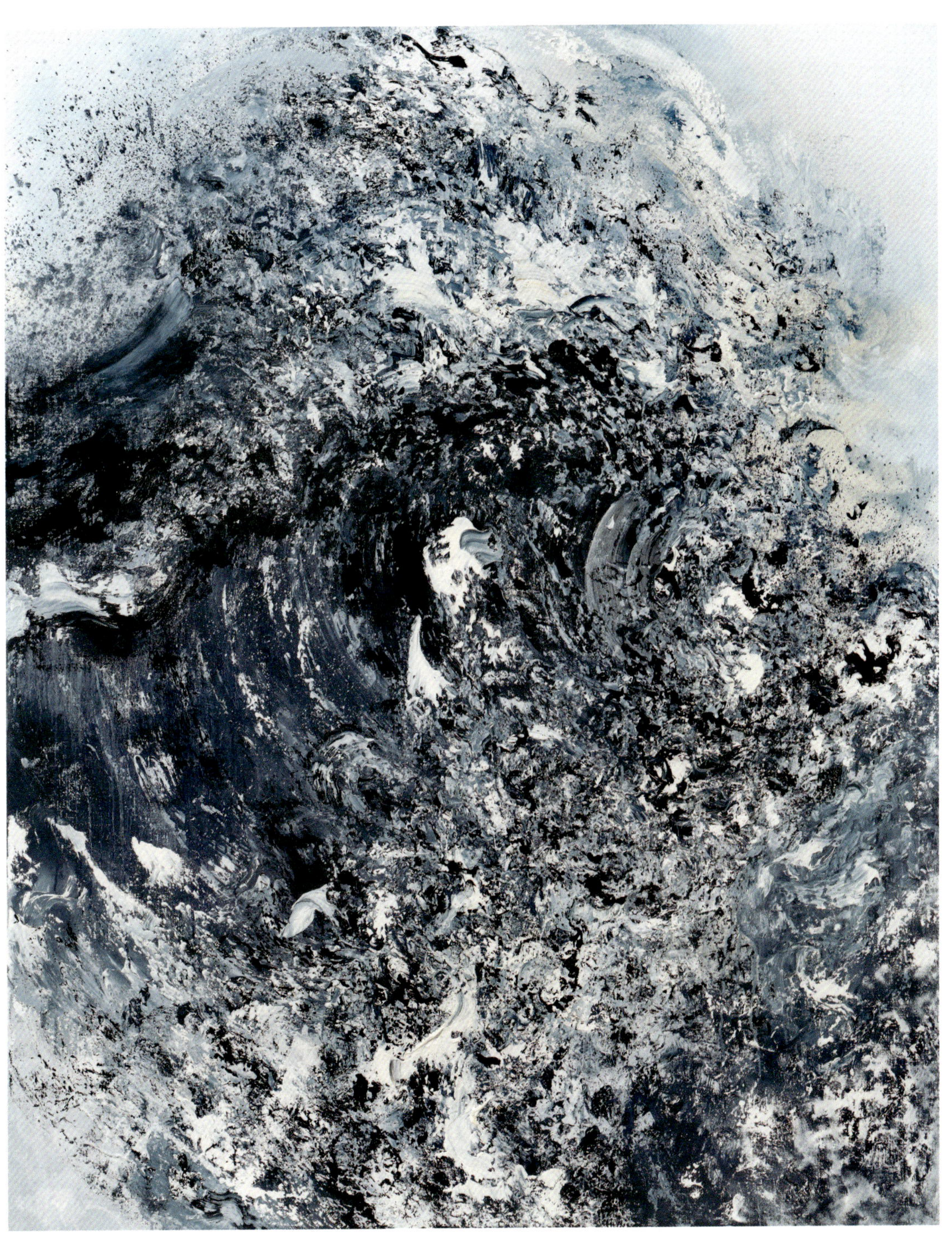

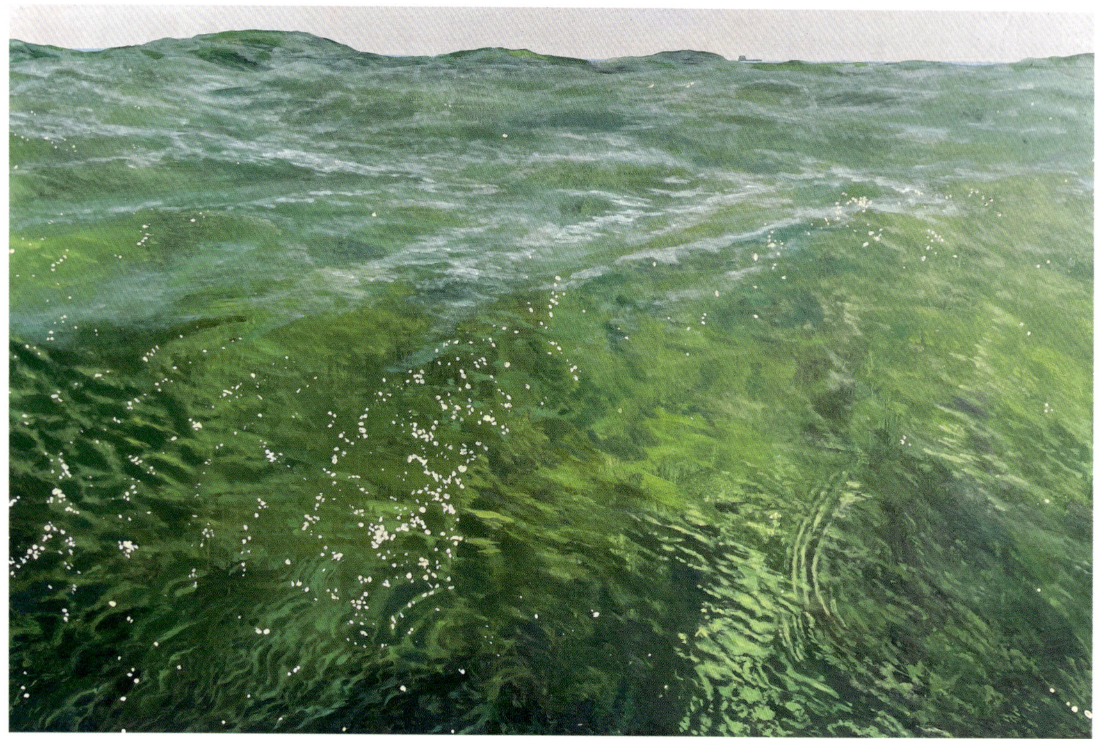

Fig. 6.3
Claire Cansick,
I Can See the Sea,
2022, oil on canvas

wrote in 2010 – the same year that she started this series – it is 'terrifying, beautiful, rapacious, embracing, never still, always hungry, always seducing, always mysterious, always there'.[2] This vast and visceral understanding of the sea permeates the series, each work a fleeting attempt to capture the portrait of a single wave. The works therefore become a lesson in ephemerality. Ultimately, the impermanence of a wave is comparable to that of a human subject in the context of planetary time. This dynamic between human life – on both a singular, vulnerable, momentary scale, and a larger, destructive, species scale – and nature is undoubtedly at play in these works. Like so much of Hambling's work, death is never far away, even though the fierce energy in each piece feels like an exuberant celebration. As is common with her work, it is hard to reconcile the fractured and far-reaching ambition of Hambling's preoccupations, as is reflected in her often-quoted refrain: 'I'm trying to paint death with as much life as I can.'[3]

While Hambling's walls of water mark a clear shift from recognisable form to barely contained abstraction, Claire Cansick's seascapes waver between the two. Painted from snapshots made whilst swimming, the field of *I Can See the Sea* (2022) is almost entirely water, the horizon line floating at the opposite edge to Hambling's works, pushing the sky out of view (fig. 6.3). Yet the effect is remarkably similar, the foreshortened perspective leaving the viewer immersed, at risk of full submersion. The canvas embodies a similar ambition to give shape and form to the potent power of the sea, to represent it in all its glory, capturing its literal and metaphorical depths. Much like Hambling, Cansick's use of colour – from flat ochres, moss green to shimmering blues – reveals a shared pursuit of a true representation of the ineffable and intangible nature of their native water. The interweaving of humankind and nature is here too, in the almost indistinguishable tanker in the distance. Such a ship is a reminder of the historical trade routes and maritime commerce that exists in the North Sea. If we were to extend the symbolism further, it could also point to the emergence of new and hugely profitable shipping routes across the Arctic that are only emerging as sea ice starts to disappear.

Margaret Mellis's driftwood assemblages interject into the dynamic of human beings and the sea in a different way. Mellis's personal history and artistic journey is rich and substantial. By the time she started creating these abstract sculptural collages from driftwood collected on walks, she was widowed and divorced, had relocated from the burgeoning scene of St Ives to rural South of France and then to the Suffolk coast, all the while searching for an authentic visual language. Just as Hambling talks of emptying herself in order to let her subject flow through her, Mellis's process of making her assemblages was intuitive and spontaneous rather than laboured. She let the materials speak for themselves, allowing the colours, shapes and textures to emerge from random arrangements on the floor. As Harriet Baker observed in her succinct and enlightening introduction to Mellis's life and work, '[t]he end announced itself'.[4] Mellis herself wrote: 'When I've got to the point where nothing can be removed or rearranged without undoing or spoiling what I have arrived at, something else has emerged out of the transfigured total.'[5]

The pieces of driftwood retained elements of their original purpose – the coloured paint of boats and industry – yet were indelibly scarred and marked by the sea. While they were undoubtedly forged with durability in mind, these man-made items were no match for the effects of water and time. Mellis's reframing of these materials could be seen as an attempt to wrest control over a more powerful force. She takes something the sea has shaped and reshapes it, ultimately inserting herself back into a larger narrative. In her 1991 assemblage, *Cloud Cuckoo Land* (fig. 6.4), Mellis has borrowed the conventional framing of a painting, and the overall effect is a similar sense of horizon line and perspectival foreshortening as Hambling's and Cansick's canvases. The collage is suggestive of a seascape, wooden supports foregrounding a body of water, the

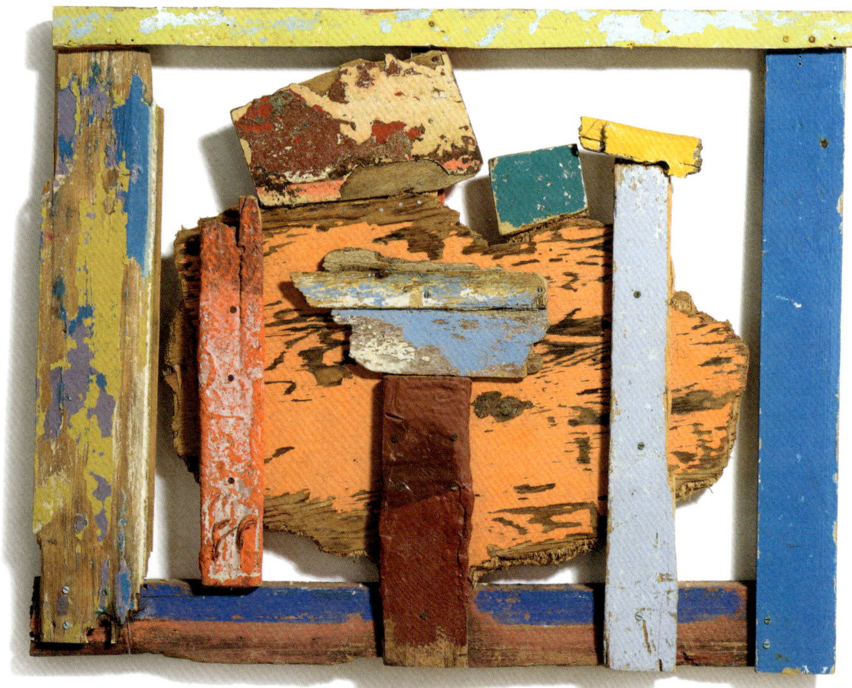

Fig. 6.4
Margaret Mellis, *Cloud Cuckoo Land*, 1991, driftwood construction. Government Art Collection

worn-away paint on wood like light hitting rippling waves, with floating vessels in the distance. The work shares a similar colour scheme; ochres and pinks interspersed with the expected sea and sky blues.

Each of these artists' work is illustrative of a specific experience, a particular interface with the sea. Each woman's bodily immersion in their natural world through seemingly meditative experiences – watching, walking, swimming – inspires an urgent need to grasp the specificity of the fleeting embodied moment. Undoubtedly there is a power in the immediacy of these works. Whether the sense of aggression that Hambling allows to flow through her onto the canvas, or the quiet intuition that Mellis funnels into collage, these pieces captivate the viewer. The parameters of the artwork become a portal, the bodily experience of the artist in nature permeating to impact upon the bodily experience of the viewer, however subtly. But it is hard to imagine that this is enough. In the context of the immensity of the climate emergency, this sharing of an individually significant – yet globally minor – experience of the natural world seems hard pressed to make a difference. Yet perhaps by recentring the importance of nature, and our place in it, these works might shift something in the viewer, even if imperceptibly.

In his 2024 book *Cosmic Connections: Poetry in the Age of Disenchantment*, Charles Taylor proposes 'we need a relation to the world, the universe, to things, forests, fields, mountains, seas ... ; where we feel ourselves addressed, and called upon to answer'.[6] While exploring poetry, not painting, his argument is compelling in relation to these artworks. The thrust of his thesis is that, post-Enlightenment, human beings have become increasingly 'disenchanted', seeing the natural world less as an awe-inspiring, wondrous cosmos that enfolds us, and rather as something observable, quantifiable and domitable.

Similar to Taylor's counter-Enlightenment entwining of humans in the natural world, cultural historian Astrida Neimanis argues a feminist positioning of our relationship to water and the sea in her essay 'Hydro-Feminism: Or, On Becoming a Body of Water'.[7] While her argument is expansive and metaphorical in part, she also is unequivocal in our literal 'wateriness'. We are both of water as well as inherent to water systems – both man-made and natural – as is the rest of the planet. Our demarcations are porous, and an 'aqueous understanding of our *interbeing*' is necessary to repair the man-made separation of humans and water, fuelled by the unnatural agendas of colonialism and capital.[8]

Neimanis's and Taylor's theoretical aggrandising of the natural world is part of a much wider shift towards 'posthumanism' in philosophical thought, which I can only touch on here. Although perhaps an oversimplification, this reconsideration of the human, nonhuman and technological worlds upsets the understanding of a human-centric hierarchy. Nevertheless, I would argue that the theoretical remains alienating to the majority, reinforcing the prioritisation of the mental over the physical plane. What is so important about the works discussed in this text is their capacity to affect their viewer's experience beyond their learned knowledge.

A 2024 article in the *Guardian* brings these ideas sharply into a quotidian focus. 'Dictionary definitions: The quest to put the human back in nature' introduces the attempts by businesswoman and environmental activist Frieda Gormley and lawyer Jessie Mond Wedd to alter the Oxford English Dictionary's definition of 'nature', which separates and opposes the natural world and humans and their creations.[9] As the article outlines, this oppositional definition is not only dangerous, in that in understanding themselves as distinct from the natural world, humans feel less responsibility to protect it, but scientifically inaccurate. The women discovered an older definition of the word, considered obsolete since 1873: 'In a wider sense, the whole of the natural world, including humans and the cosmos', which has now been brought back into contemporary definition by

OED lexicographers.[10] It is this dynamic – or conversation, as Taylor proposes – with the natural world that is crucial to impacting our relationship to the climate emergency. If we can experience a felt immediacy and urgency in regard to the natural world, we might be inspired to translate this to a global scale. The artworks mentioned in this text are not theoretical or logical arguments for a defence of the sea, but rather are experiential moments of a single human's embodied understanding that she is part of a larger order. As Taylor explains:

> *[W]hat we seek to recover contact with is no longer a cosmos, but Nature; that is, the complex, interwoven, and vulnerable order of living beings on our planet. Many of us seek to recover a sense of belonging to, of membership in this larger order, to be nourished by it, and in solidarity with it, in the face of the terrible destruction which now threatens it.*[11]

While it is hard to know whether Mellis was conscious of the context of humankind's extreme negative impact on the environment when making her work – certainly the hole in the ozone layer was a cultural conversation in the 1980s – Hambling and Cansick have both been explicit in their concern for the climate emergency. Nevertheless, it is clear that all three artists show an awareness of the symbiotic relationship between humans and the sea. In successfully conveying their own individually experienced 'tipping points', Hambling, Cansick and Mellis both stymie and embrace their connection to this 'larger order'. In the context of the theme *Can the Seas Survive Us?*, these works cannot be mistaken for gentle seascapes but should be read as urgent missives to reveal our place in a natural system. The sea is not something we bear witness to; it is something to which we are inextricably linked. While Hambling, Cansick and Mellis may not be overt in their activism, the works discussed hold a soft power. Rather than raising awareness or coercing concern through didactic means, these artists urge the viewer. Through creating a shared experience of a moment of being connected to nature, they convince the viewer. In other words, they *move* us to care.

The vitality of these pieces does not mourn what has been done wrong, nor map a pragmatic commitment to change. Instead, they reveal an ability to stop where we are and truly see our reality. In doing so, we might better understand the holism of our planet, and our subsequent responsibility to it. The tides are turning, and we cannot hold them back.

'SELKIE SONG'
HARUN MORRISON

Notes for 'Selkie Song'

'Selkies', also known as 'the seal people', are mythical creatures capable of transforming from seals into humans. This unique ability to shift between the sea and the land has fascinated various cultures, leading to many stories and legends. These stories often explore themes of freedom, identity and the mysterious bond between humans and nature. They feature prominently in the oral traditions and mythology of various cultures, especially those of Celtic and Norse origin. The term 'selkie' derives from the Scots word for 'seal', and is also spelled as silkies, sylkies or selchies. The narrative for this song was shared with me by Copenhagen-based artist Marie Kølbæk Iversen.

Selkies also function symbolically as blurring the human and nonhuman; as a go-between their being is transgressive of boundaries between sea and land, seal and human. Many selkie tales involve tragic encounters between human and selkie, which can be read as a warning for humans against these border crossings.

'SELKIE SONG'

Melody: 'Sømandens himmerige' (originally 'Fiddlers' Green') by John Conolly, 1966
Lyrics commissioned by Hosting Lands, 2024

[1] As I walked / by the cliffside / one evening / so bright,
I stared / at the cliffs / could not trust my sight.
I caught a group of Selkies / curled in the sun.
I was drawn toward them, / I could hear their song.

Refr. I wifed a Selki / till she fled from me
Perhaps you'd say I stole her
I never stole her heart, / it's buried in sea salt
O what a choice – young ones / or the water.

[2] Seeing me, they dived / away / seizing their skins,
They all escaped but one, / and so began my sins.
Where her skin goes she went / and she became my wife,
I hid her hide, I lay with her, she bore four lives.

Refr. I wifed a Selki / till she fled from me ...

[3] By now my hair was grey / and no fish caught today,
Our kids found her old seal pelt buried in some hay,
They took it to their mother, asking 'what is this?'
They told me, she replied: 'It's something I have missed.'

Refr. I wifed a Selki / till she fled from me ...

[4] My empty boat rowed to shore / I clocked my wife and kids,
She kissed each one, then jumped into her skin,
Jumping out / my skiff, I fell / she cried, 'please, treat them well',
As her legs turned to tail she dived in / the stormy swell.

Refr. I wifed a Selki / till she fled from me ...

[5] Now, I'm lonesome / I try to soothe our weeping kids,
Did I commit a crime? I ignored the myths.
On Sundays we return / to the windy shore,
We cry into the waves for who we've lost and more.

Refr. I wifed a Selki / till she fled from me ...

A SEA OF RESILIENCE: OCEANIC (IN)VISIBILITY

KAREN JACOBS

The term 'Oceanic' in this chapter's title refers to Oceania, a region that is defined by the Pacific Ocean. Covering one-third of the world's surface, the Pacific Ocean (known in Oceania by different local names) is larger than the landmass of all of the continents grouped together. Yet while vast in terms of ocean space, Oceania is also considered the smallest in surface land area and, after Antarctica, the least populated. Consequently, Euro-American conceptions of Oceania have historically invoked the dialectic of centre and periphery to differentiate Oceania from the West. The emphasis on Oceania's seemingly peripheral value within European imperialism has been visually captured in the cartographic representations of the region produced from the sixteenth century onwards. These maps draw attention to small islands in a vast ocean. However, in the 1990s Tongan scholar Epeli Hau'ofa put forward a vision for Oceania in response to the Euro-American tendency to divide the region into clearly delineated and separate areas resulting from colonialism and developmentalism. Rather than a series of small, isolated entities, in his essay 'Our Sea of Islands' (1994) Hau'ofa proposed to view Oceania as an assemblage of islands connected by the ocean; an ocean that acts as a linking pathway rather than a separating boundary.[1] This is a pathway that is reclaimed by people today by celebrating the impressive navigational skills that their ancestors used when settling on the islands. These ancestors considered their homeland 'comprised not only land surfaces but the surrounding ocean as far as they could traverse and exploit it'.[2] By considering the ocean space as part of their homeland, and by emphasising historic and ongoing mobility in the region, Hau'ofa defies the label of isolated periphery and demonstrates that 'their world was anything but tiny'.[3]

In their representations of Oceania, other scholars from the region also emphasise the common relationship with, and connection to, the ocean for the Oceanic people living in the region.[4] Therefore, the concept of 'Moana' meaning 'ocean' in some, but not all, Pacific languages, was proposed by Chitham, Māhina-Tuai and Skinner in their book *Crafting Aotearoa*, to be used to refer to the region even by 'island nations that do not have the word Moana in their languages', as it privileges Moana or Indigenous perspectives and aims to challenge the European origins of the name Oceania.[5] The term Océanie (Oceania) was first introduced in 1815 by French geographer Adrien-Hubert Brué (1786–1832), who drew on the term Océanique that had been coined in 1804 by French geographer and cartographer Edme Mentelle (1730–1816) and French journalist and geographer Conrad Malte-Brun (1775–1826) as equivalent to Afrique (Africa) and Amérique (America). In 1832, French explorer Jules-Sébastie-César Dumont D'Urville (1790–1842) endorsed the term Océanie in his well-known essay, 'Sur les îles du Grand Océan', in which he divided the region further into Polynesia, Micronesia and Melanesia (in which he included Australia) and the term Oceania became widely used.[6]

This division of Oceania into separated regions not only informed subsequent Euro-American representations that highlighted Oceania's isolation, but it equally informed the choice by post-Second World War nuclear powers to conduct atomic testing in the Pacific Ocean, with disastrous consequences for the ocean and its inhabitants. Today, the same reasoning informs the response by Western nations to the climate crisis.[7] The Pacific Ocean is threatened by global warming and associated extreme weather events, rising sea levels, plastic pollution, coastal erosion and species extinction. However, as Jaimey Hamilton Faris has observed, the problem of climate change is not simply the submersion of the islands but the equal submergence and invisibility of the communities who reside there.[8] While not denying the impact of global issues – which are often the result of unequal power relations – on the region and its people, this chapter focuses on a range of artists based in Oceania who challenge this invisibility and emphasise resilience.

A DYING SEA: NUCLEAR TESTING AND PLASTIC POLLUTION

At 9am on 1 July 1946, an atomic bomb exploded at Pikinni (at the time known as Bikini, a coral atoll in the Marshall Islands in the Pacific Ocean), with an energy yield of 23,000 tonnes of TNT. The media covered the event widely through images of mushroom clouds blooming out of the ocean, expanding upwards and outwards. These abstract, almost innocent images literally clouded the enormous impact of the bomb on the region, which marked 'a new age in both ecology and imperialism'.[9] Four days later, on 5 July, Paris-based clothing designer Louis Réard (1897–1984) named his new design of a two-piece swimsuit after the test site. Naming the garment after the nuclear testing site was a way to signal that both inventions – the garment as well as the scientific impact of the atomic test – were forward-looking, excited for a modern future. However, the garment's name connection to the Oceanic testing ground gradually faded. Pikinni/Bikini Atoll was moved to the background; rendered invisible. As Teresia K. Teaiwa has written: 'By drawing attention to a sexualized and supposedly depoliticized female body, the bikini distracts from the colonial and highly political origins of its name.'[10]

Clothing seems to be a common thread. It was the fact that an American friend, and collector of antique Hawaiian shirts, told Tasmanian artist Pam Debenham (b.1955) that the rarest shirt from the 1950s was one produced in celebration of the US nuclear testing on Pikinni/Bikini Atoll, that inspired her 1984 work *No Nukes in the Pacific* (fig. 8.1). The screenprint features a figure wearing a Hawaiian shirt with depictions of mushroom clouds amid palm trees and a boat named the *Pacific Peacemaker*. Each time the scene is repeated on the fabric it is topped

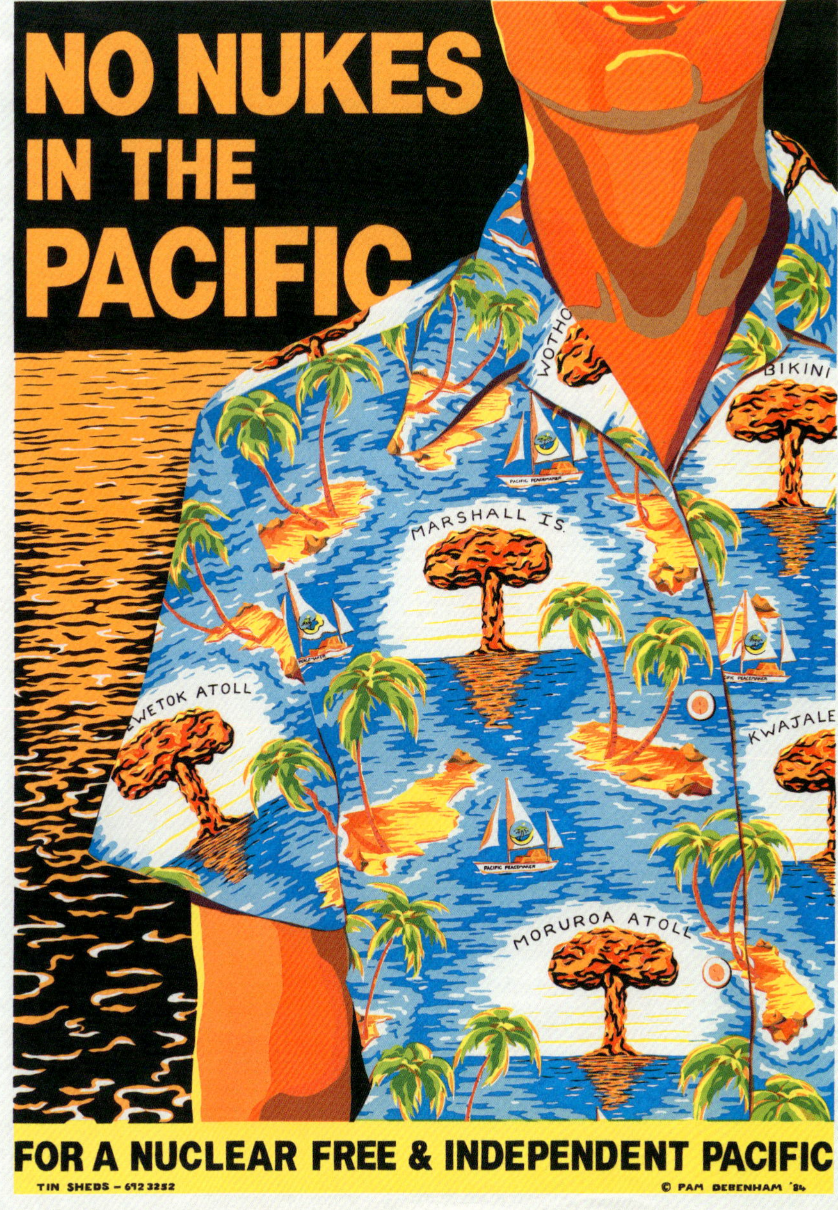

Fig. 8.1
Pam Debenham, Tin Sheds Posters and Sydney University Art Workshop, *No Nukes in the Pacific*, 1984, screenprint in coloured inks. National Gallery of Australia, Canberra

with the name of a different nuclear testing location in the Pacific Ocean. A clear message of protest and peace, *No Nukes in the Pacific* emphasises the ubiquity of places in the Pacific Ocean where nuclear bombs exploded and makes a demand to stop atomic testing.[11]

It is estimated that more than 315 nuclear tests were conducted in the Pacific Ocean between 1946 and 1996 by the US, the UK and France.[12] However, figures vary, and the exact number is unknown as information remains classified. Nuclear testing marked a new age of imperialism as post-Second World War colonial powers used their territories as nuclear testing sites. Evidently the impact of nuclear testing was immense, and the effects are still felt today. In addition to their sheer blast power, nuclear weapons also eradicated life through long-lasting nuclear fallout and radiation leaks from underground test sites or radioactive waste stores. It is not just the ecosystem that has been irreversibly damaged; local people were afflicted with cancers and genetic illnesses related to the radioactivity.[13] Evacuation of Islanders had disastrous consequences, rupturing the connection of residents to their homeland. Inhabitants of Pikinni Atoll and Enewetak Atoll were relocated by the US, many against their will or in the belief that their relocation would only last a few years. Pikinni Atoll remains uninhabitable due to dangerous levels of radiation and contaminated water. Enewetak people could return after a belated partial clean-up of radioactive waste in the 1970s, but nuclear fallout and the removal of the topsoil during the clean up made the land unsustainable meaning residents were reliant on imported food.[14] As Nic Maclellan points out: 'Seeking "empty" spaces, nuclear powers chose to conduct Cold War programs of nuclear testing in the deserts of central Australia or the isolated atolls of the central and south Pacific. But these regions were not "terra nullius".'[15] The people who lived in these spaces, who were evacuated, who became ill or who lived under the threat of radiation poisoning, became invisible.

In her artwork *See Reverse Side* (2017), Jane Chang Mi (b.1978) draws on the photographic archive of Operation Hardtack I in the Marshall Islands, a series of 35 nuclear tests conducted by the US in 1958. The photographic archive is considerable because photography was a way of recording scientific results, including explosion and blast forces. As a result of this scientific focus, out of the hundreds of photographs taken of nuclear testing in the Marshall Islands, only a few depict Marshallese residents. Mi found 13 declassified photographs in the National Archives and Record Administration (NARA) showing the way of life that was destroyed. She chose to redraw the archival photographs to refocus attention on them, as such demonstrating the human realities of cultural displacement and land loss.[16] The image illustrated shows one of these redrawn archival cards (fig. 8.2). The juxtaposition between recto and verso, and between image and text, is significant. The front of the card shows a drawing of people from the Marshall Islands, an exact copy of the photograph, with the text: 'Natives and Native villages on Bikini Atoll Marshall Islands site of the 1st peace time test of the atomic bomb against a massed fleet. The pictures are the

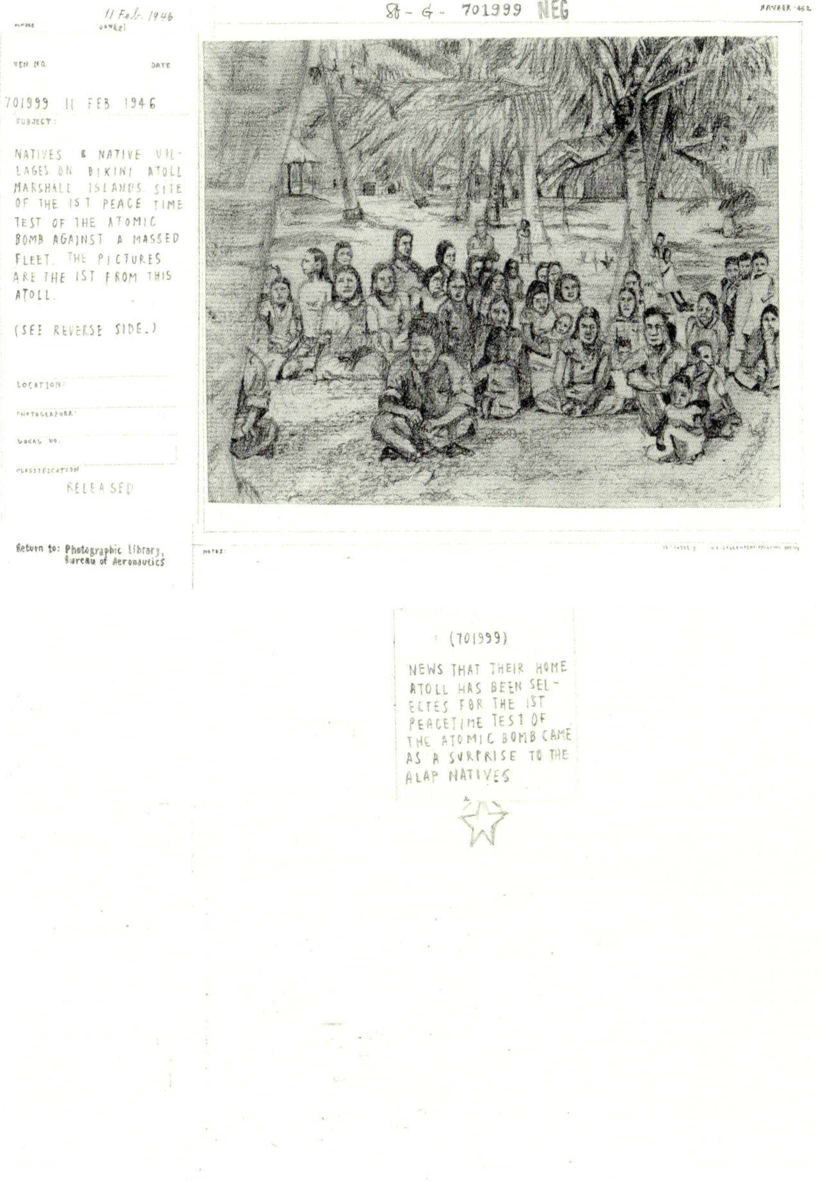

Fig. 8.2
Jane Chang Mi,
See Reverse Side,
2017, pencil on paper
(recto and verso)

1st from this atoll. (See reverse side).' The other side of the card reads: 'News that their home atoll has been selected for the 1st peacetime test of the atomic bomb came as a surprise to the Alap Natives.' Mi's artwork draws attention to the tension between history and memory of the atomic era by redirecting the focus of the atomic experience away from science to the people who lived there. It presents an unnerving juxtaposition of extraordinary authority and ordinary complacency.

In the late 1970s an attempt of remediation was undertaken at Enewetak Atoll when highly contaminated topsoil was partially removed, dumped in a crater caused by nuclear testing and buried under a concrete dome structure on the atoll's Runit Island. The concrete dome that contains 73,000 square metres of radioactive waste plays a central role in the video poem *Anointed* (2018) created by Kathy Jetñil-Kijiner (b.1989) with filmmaker Dan Lin (b.1973). Born in the Marshall Islands, Jetñil-Kijiner's work as artist, poet, scholar and curator focuses on threats faced by Marshallese Islanders such as climate injustice and forced migration, American nuclear testing and militarism. *Anointed* is a video of a poem recital by Jetñil-Kijiner while she visits the concrete dome on Runit Island. Images of Jetñil-Kijiner sailing to the atoll are interspersed with images of playing children, breadfruit and island scenes, as well as archival footage of nuclear bomb detonations filmed by the US military. 'You were a whole island, once', says Jetñil-Kijiner. By asking 'Who would have us forget that you were once green globes of fruit, pandanus roots and whispers of canoes? Who knows the stories of the life you led before?', she draws attention to previous life on the island as opposed to the empty, desolate scenes of destruction and devastation. She demonstrates that people once lived there but now all that is left is 'a concrete shell that houses death'. The video ends with Jetñil-Kijiner wearing the *jaki-ed* – a Marshallese finely woven clothing mat – walking barefoot on the concrete dome to place white coral stones in the centre – a Marshallese funerary custom.[17]

Referring to the atoll Moruroa (also spelled Mururoa) in the Tuamotu archipelago that was at the centre of France's nuclear testing operation, the Pacific Sisters created *Mururoa* (2018; fig. 8.3). *Mururoa* (pronouns we/you/they) is part of three *aitu* (avatars) that the Aotearoa New Zealand-based artist collective created to bring to life the songs sung by Sister Henry Ah-Foo Taripo (b.1965). *Mururoa* refers to the song 'Moruroa' that protests the damage and destruction caused by France's era of nuclear testing. The song emphasises Pacific unity against the nuclear age. *Mururoa* combines separate adornments made by Pacific Sisters Feeonaa Clifton (b.1972), Ani O'Neill (b.1971), Lisa Reihana (b.1964) and Suzanne Tamaki (b.1962). Each piece refers to the impact of nuclear testing and combines

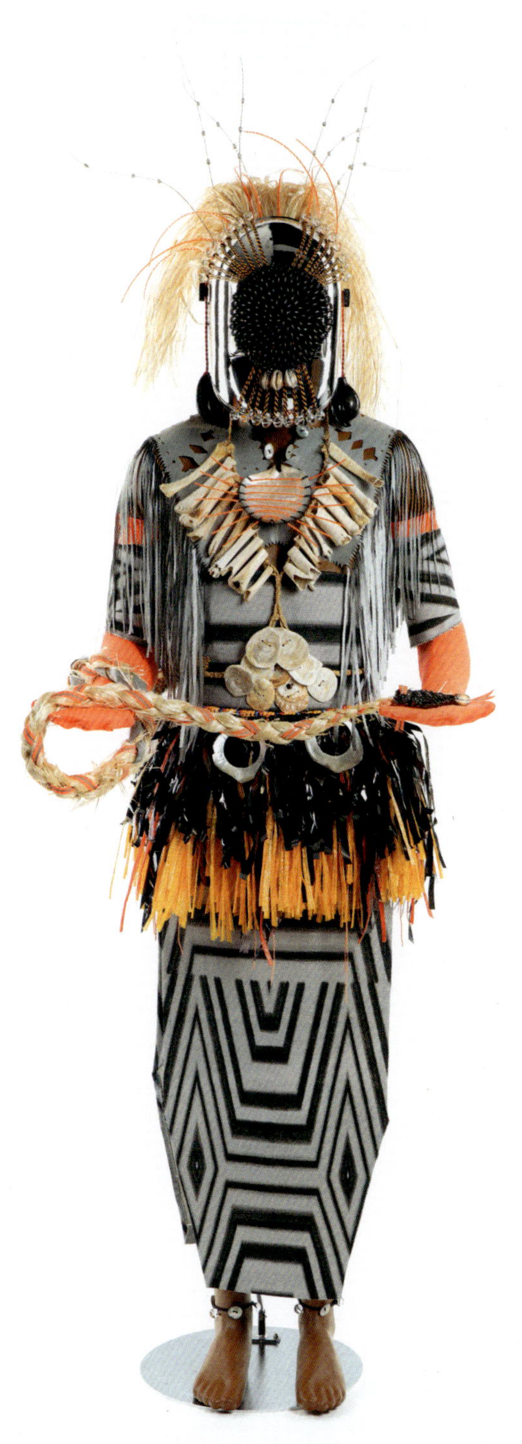

Fig. 8.3
Pacific Sisters,
Mururoa, 2018,
textile, plastic,
twill and bone.
Auckland War
Memorial Museum
Tāmaki Paenga Hira

the Pacific Sisters' artistic practice of using different innovative materials. The *Cyclops in Radioactive Paradise* mask is made of materials such as an upcycled electroplated welding mask, cowrie shells, glass beads, Job's tear seeds, plastic raffia and coconut leaf midribs. The ripped layered stockings that the figure wears are fittingly called *Meltdown Skin*. The anklets made of strung glass beads and pearl shell buttons are entitled *Radiation Falling.*[18] Together the 11 pieces of adornment constitute a post-apocalyptic figure who reminds the audience of the ongoing legacy of nuclear fallout and the impact of rendering people invisible.

It is not just the legacy of nuclear testing that impacts the Pacific Ocean, but other anthropogenic waste forms also pose a significant threat to the ocean. A 2017 United Nations report stated that if single-use plastic items such as bags and bottles were not stopped, by 2050 our oceans will contain more plastic than fish.[19] Artist George Nuku (b.1964), of Māori (Ngati Kahungunu/Ngati Tūwharetoa) and German/Scottish descent, calls attention to the plastic problem in the Pacific Ocean in his *Bottled Ocean* series. The *Bottled Ocean* works are made of recycled plastic bottles that have been turned into sea creatures, corals or canoes. Each iteration of the artwork provides a projection 100 years into the future, when plastic has overpowered life on Earth, a process that has already started if we consider the Great Pacific Garbage Patch, an island of non-biodegradable marine debris that floats in the Pacific Ocean. Following reality, Nuku creates water worlds and a grim vision for the future. However, his aim is to encourage audiences to revalue the beauty of plastic by showing its ancestral relations. He

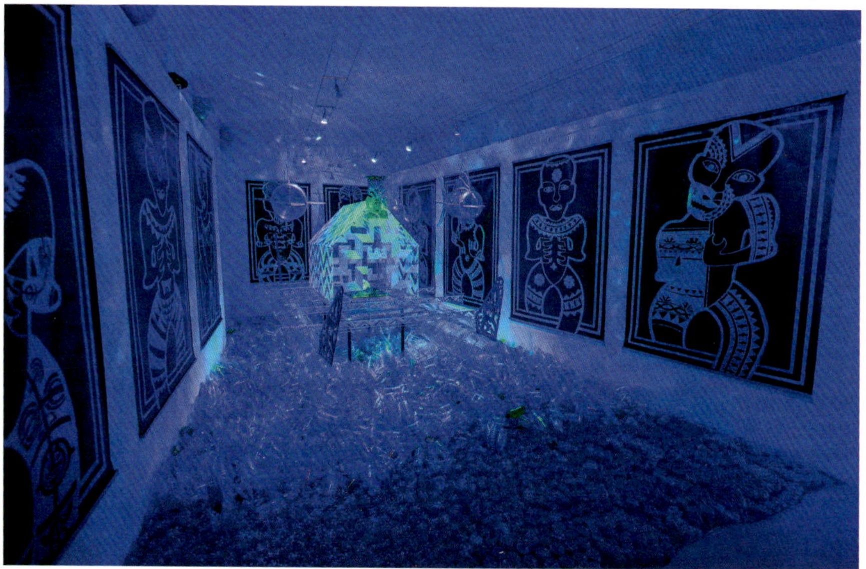

Fig. 8.4
George Nuku,
Bottled Ocean 2124,
2024, reclaimed
plastic

draws on the link between petroleum and plastic and wants to demonstrate that plastic is not the cause of death but a source of life – if we treat it well.[20] *Bottled Ocean 2124* (fig. 8.4) was made on location at the Musée du Masque de Fer et du Fort Royal in Cannes, France (17 May–17 November 2024). Plastic bottles are used to depict corals and sea creatures that accompany a plastic *waka hourua* on a sea of plastic bottles. A *waka hourua* is a double-hulled canoe, the type that was used to settle in Aotearoa New Zealand during ancestral voyages. The *waka* is at the basis of Māori *whakapapa* (genealogy) as people recite the name of the *waka* on which their relatives travelled to Aotearoa New Zealand during the first settlement. As such, Nuku emphasises mobility and unity in the region in the face of global plastic pollution.

'Ghost net' artworks equally bring to life discarded items from the oceans, in this instance, fishing nets. Various Australian Aboriginal and Torres Strait Islander communities are now transforming ghost nets – abandoned fishing nets that have washed ashore – into works of art to raise awareness of the devastating threat to the environment that these nets cause. The nets are turned into turtles, fish, sharks, frigate birds and other creatures that are often trapped in them.[21] *Bee Dee* (2016) is a ghost net turtle sculpture created by Florence Gutchen (b.1961) from the Erub Islands in the Torres Straits, Australia (fig. 8.5). It is entirely made from blue, green, yellow, black and white reclaimed plastic fishing nets. Transforming something dead and dark into something beautiful and hopeful emphasises resilience rather than despair and isolation. Yet it also

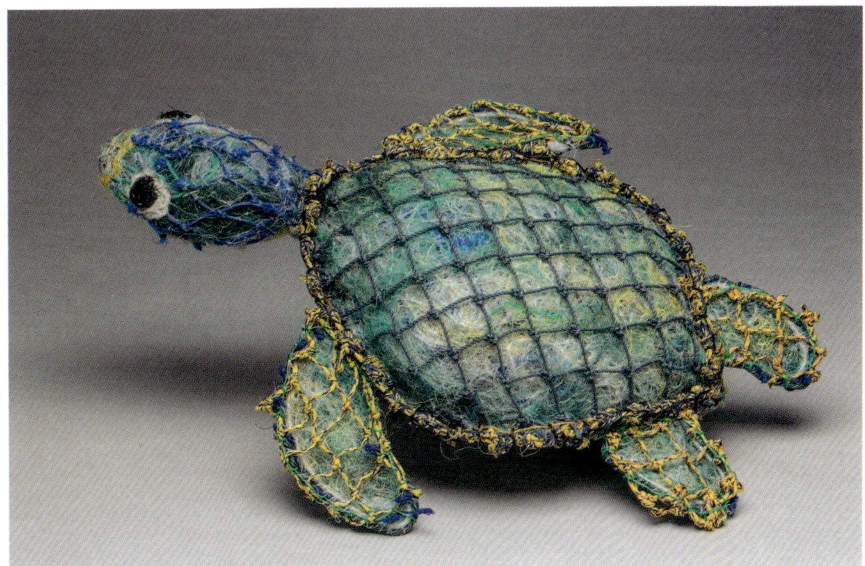

Fig. 8.5
Florence Gutchen, *Bee Dee*, 2016, reclaimed plastic fishing nets. Museum of Archaeology and Anthropology, Cambridge

draws attention to the problem of waste in a direct and visual way that engages the public more than statistics ever could.

A RISING SEA: CLIMATE CHANGE

Many of the Pacific Islands are low-lying atolls, which are impacted by the current high pace of sea level rise and threatened by flooding, coastal erosion, coral degradation, loss of groundwater sources and surges in extreme weather phenomena.[22] While Oceania has contributed the least but suffered the most from climate change, the regular focus on isolated, sinking, drowning small islands in Oceania tends to victimise people in the region to a degree that they lose agency.[23] Katerina Teaiwa's 2018 essay 'Our Rising Sea of Islands' expands Hau'ofa's perspective on Oceania to the context of climate change.[24] Teaiwa's 'rising sea' refers not only to sea level rise, but also to the rising visibility of activism as Oceanic people make clear that 'the swallowing of such tiny islets of land by the ocean are the precursors of a much larger problem'.[25]

The video *Holding On* (2015) by Sāmoan artist Angela Tiatia (b.1973) was shot on Funafuti in Tuvalu, a Pacific atoll nation urgently experiencing the flooding

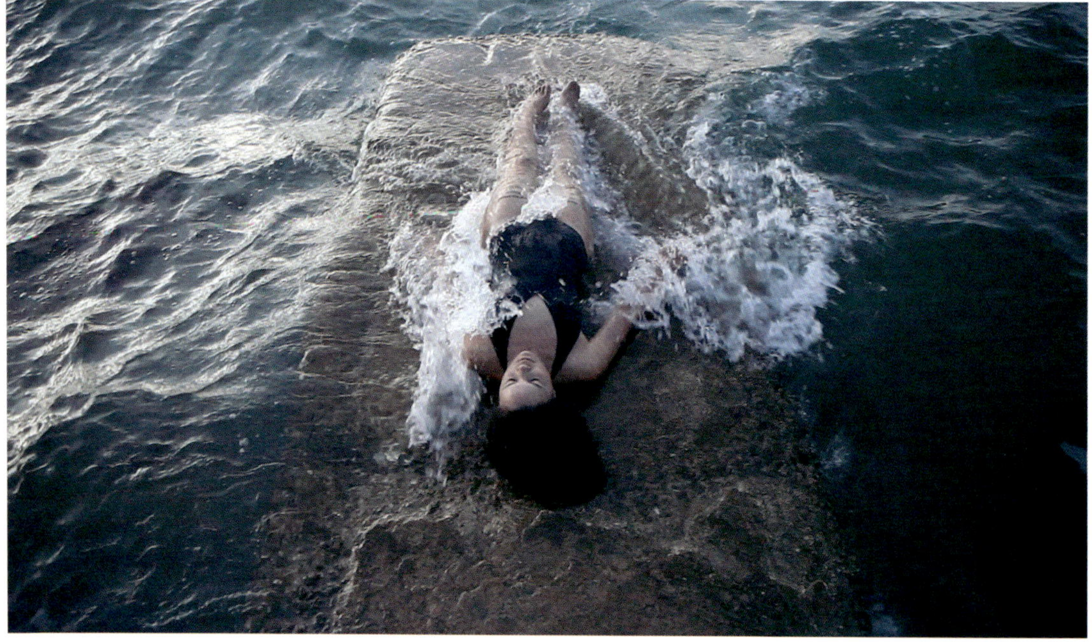

Fig. 8.6
Angela Tiatia, *Holding On*, 2015, film still

effects of climate change and global warming. The video shows Tiatia lying on a concrete slab as the surrounding surging ocean tides rhythmically lap and wash over her (fig. 8.6). At the danger of being washed away, she holds on, referring to people in Tuvalu holding on to their lands. She draws attention to the devastating impact of rising sea levels, as, in her words, 'water is the giver of live, but also the great remover of life'.[26]

Similar ideas are expressed in the video poem *Rise* (2018) by Kathy Jetñil-Kijiner and Aka Niviâna. The work by an artist from the Marshall Islands (Jetñil-Kijiner) and from Kalaallit Nunaat (Greenland; Niviâna) was commissioned by environmental activist organisation 350.org. As 'Sister from ocean and sand' and 'Sister from ice and snow', both artists express their relation to the environment, one where icecaps are melting, another where the water is rising. They meet in Kalaallit Nunaat, where Niviana asks Jetñil-Kijiner:

> *Sister of ocean and sand,*
> *Can you see our glaciers groaning*
> *with the weight of the world's heat?*
> *I wait for you, here,*
> *on the land of my ancestors heart heavy with a thirst*
> *for solutions*
> *as I watch this land*
> *change*
> *while the World remains silent.*

Both women address more than the impact of melting/rising water in a physical sense; they discuss the drowning or the silencing of their communities in an expression of solidarity between Indigenous people dealing with a global problem.[27]

An artist who has referenced the threat of the rising ocean across her creative repertoire is multimedia and performance artist Yuki Kihara (b.1975). Her work deals with transnationalism, gender identity and the ongoing impact of climate change on ocean habitats around the islands of Sāmoa. Her 2013 series *Where do we come from? What are we? Where are we going?* consists of photographs taken at places damaged by the tsunami that hit Sāmoa in 2009 and the fallout from Cyclone Evan in 2012 (fig. 8.7). Wearing a Victorian-style mourning dress, Kihara, dressed as her alter ego Salome, stands near the remnants of a series of buildings that were ravaged by the impact of these extreme weather events. By the time the photographs were taken, the sites had been cleaned up

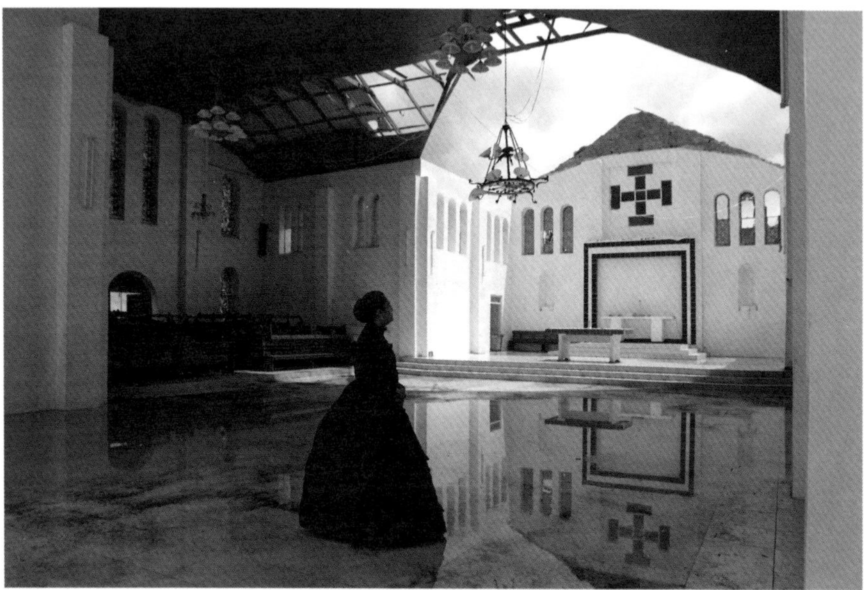

Fig. 8.7
Yuki Kihara,
Agelu i Tausi Catholic Church After Cyclone Evan, 2013, C-print

and the resulting eerie emptiness around the ruins emphasises the ongoing effects of climate disaster. Kihara's choice of dress expresses grief and loss, but the reference to Victorian costume equally references climate change as a colonial act.[28]

The threat of climate change is also a theme that is unpacked in Kihara's work *Paradise Camp,* which drew positive reviews and large audiences at the 59th Venice Biennale in 2022.[29] As the first Oceanic, Asian and Fa'afafine (third gender) artist to represent Aotearoa New Zealand at the Biennale, Kihara's appointment highlights Aotearoa New Zealand's acknowledgment of its long-standing engagement with the Pacific region, including its impact on Sāmoan gender identities.[30] Challenging the notion of the Pacific as a paradise and claiming that there is more camp in the paintings of French post-Impressionist artist Paul Gauguin (1848–1903) than initially meets the eye, *Paradise Camp* presents an immersive installation that contests gender conceptions and addresses climate disaster and a threatened ocean.

Since its showing in Venice in 2022, *Paradise Camp* toured to Powerhouse Museum, Gadigal land Sydney in 2023. In 2024 it was on view at the Saletoga Sands Resort and Spa on Upolu Island, Sāmoa, before opening at the Sainsbury Centre, University of East Anglia, in 2025. While *Paradise Camp* has been expanded and adapted for each location, each iteration consisted of three core elements: 12 large-scale photographs, two videos and the *Vārchive*. Its adaption for the Sainsbury Centre is discussed between Kihara and the exhibition's curator, Tania Moore, in the conversation following this chapter.

The 12 tableau photographs reimagine and reconstruct Gauguin's paintings created during his time in Tahiti and the Marquesas Islands between 1891 and 1903. Eleven photographs were shot on location in Sāmoa, with a large local cast and crew. While it was always assumed that Gauguin painted young Tahitian women, Kihara's extensive research uncovered that he drew on photographs of Sāmoans taken by Aotearoa New Zealand colonial photographers, such as Thomas Andrew (1855–1939), who had a studio in Sāmoa between 1891 and 1939. Kihara revealed that Gauguin most likely came across these photographs during his visit to Aotearoa New Zealand in August 1895 and she found Gauguin's signature in the Auckland Art Gallery Toi o Tāmaki visitor book.

Following this research, Kihara's photographs 'upcycle' Gauguin's paintings. The models are Sāmoan members of the Fa'afafine and Fa'atama communities, two of the four culturally recognised gender identities in Sāmoa. The term Fa'afafine – literally, 'in the manner of a woman' – refers to people who were assigned male at birth who identify as female. The masculine counterpart to this is Fa'atama – literally, 'in the manner of a boy'. The colonisation of Sāmoa and its emphasis on strict binary gender identities impacted Fa'afafine and Fa'atama communities and resulted in ongoing discrimination.[31] Fa'afafine and Fa'atama models pose in lush surroundings, adopting the same attitudes seen in Gauguin's paintings, the colour of their outfits arranged to represent the rainbow colours. These photographs are shown against a wallpaper image of the ocean taken from the shoreline hit by the 2009 tsunami. The paradise portrayed in Gauguin's paintings is impacted by extreme weather events caused by climate change as well as sexual discrimination.

The twelfth photograph, *Paul Gauguin with a Hat (after Gauguin)* (2020), was made in Aotearoa New Zealand after the artist herself took on Gauguin's persona using prosthetics, costume, moustache and wig. Kihara also features in this persona in one of two videos that are included in *Paradise Camp*. The video shows Kihara interviewing Gauguin, also played by Kihara. In the interview, Kihara tells Gauguin how she discovered that he drew on colonial photography of Sāmoan people. The role reversal allows Kihara to command her own space within art history. In the second single-channel video entitled *First Impressions: Gauguin*, members of the Fa'afafine and Fa'atama community return Gauguin's gaze by commenting on his paintings.

The *Vārchive*, an assemblage of archival material, historic and contemporary photos, demonstrates the extensive research process that underlies the artwork. The evidence of Gauguin's encounter with Andrew's photographs is provided amid images of gender-changing fish, news clippings and the artist's passports showing her different gender according to nationality. Kihara collected these contemporary and archival images to create a sense of *Vā*, indeed the *Vā* in *Vārchive*, a term coined by Kihara. In Sāmoa the concept of *Vā* is 'the space between, the betweenness, not empty space, not space that separates but space that relates'.[32] The *Vārchive* brings together seemingly divergent and isolated themes in order to draw attention to the needs of her community. Through the assemblage of images, Kihara links sexual discrimination resulting from colonialism to a threatened ocean resulting from the ongoing effects of colonialism. This was reality when some Fa'afafine and Fa'atama community members, who were banished from their families, faced isolation and did not have a family to return to after the tsunami destroyed their homes.[33] Indeed, Kihara demonstrates that these 'can no longer be seen as isolated incidents but as a vast interconnected "sea of islands"'.[34]

EPILOGUE

> *Oceania is vast, Oceania is expanding, Oceania is hospitable and generous, Oceania is humanity rising from the depths of brine and regions of fire deeper still, Oceania is us. We are the sea, we are the ocean, we must wake up to this ancient truth. ... We must not allow anyone to belittle us again and take away our freedom.*[35]

Global currents and tides affect Moana Oceania. The artists discussed in this chapter demonstrate how the ocean is a relational entity. Their artworks play a role in counteracting the regular focus on isolated, contaminated, drowning small islands in Oceania, which victimises people in the region, and demonstrate resilience, resolution and creative adaptability. Drawing on ancestral connections to the ocean, their Oceanic stories express mourning, anger and pain, but equally articulate resourcefulness and resilience without minimising the impact and hardships endured through successive generations. Their artworks are calls to radical political change and are testaments to the power of collective action.

THE EVOLUTION OF *PARADISE CAMP*

YUKI KIHARA IN CONVERSATION WITH TANIA MOORE

Yuki Kihara is an interdisciplinary artist of Japanese and Sāmoan descent. Throughout her practice she engages with European art histories and archives to question stereotypical representations of the Pacific. She created *Paradise Camp* for the 59th Venice Biennale in 2022, in which she reconceived Paul Gauguin's paintings of Tahiti and its people. She was selected to represent New Zealand at the Biennale and was the first Oceanic, Asian and Fa'afafine artist to do so. Fa'afafine is a third gender culturally recognised in Sāmoa, translating to 'in the manner of a woman'. Kihara represents the Fa'afafine community in her work and collaborates with the people of this community, engaging them in, amongst other things, climate change awareness workshops. Kihara is deeply mindful of the impact of climate change and the rising sea levels on Sāmoa and the wider Pacific, whose communities are amongst the most at risk and yet have the least impact on the environment. *Paradise Camp* is featured as part of the Sainsbury Centre's *Can the Seas Survive Us?* season. For the exhibition, Kihara conceived a new body of work exposing how Charles Darwin (1809–1882) manipulated his findings on queer narratives in nature to conform with the heteronormative values at the time.[1] Here, Kihara talks to the exhibition's curator, Tania Moore, about the evolution of this work.

[TM] You describe *Paradise Camp* as 'upcycling' Gauguin's paintings. Can you describe this process and the resulting body of work?

[YK] In the context of sustainability, upcycling means that you take something that has been discarded and breathe new life into it through creativity, for a sustainable environment. In the context of my work, the sustainability I refer to is the cultural environment that shapes and establishes norms and rules of behaviour and underscores what is considered acceptable or unacceptable. I see my process of upcycling Gauguin's paintings, by inserting new narratives that speak to the Indigenous queer experience, as a way of reducing the 'landfill' of art that perpetuates false myths about Indigenous peoples in the Moana Pacific and that is further amplified by the art market and curatorial narratives of museum exhibitions (fig. 9.1).

As Pacific writers Albert Wendt, Reina Whaitiri and Robert Sullivan have observed:

> *While the area known as Polynesia is indeed incredibly beautiful, it is also home to many thousands of people who have learned over the centuries to survive extraordinary hardships. The romantic ideas and images held by outsiders about the Pacific have plagued our people since first contact; and breaking away from*

Fig. 9.1
Yuki Kihara,
Two Fa'afafine (after Gauguin), 2020,
C-print

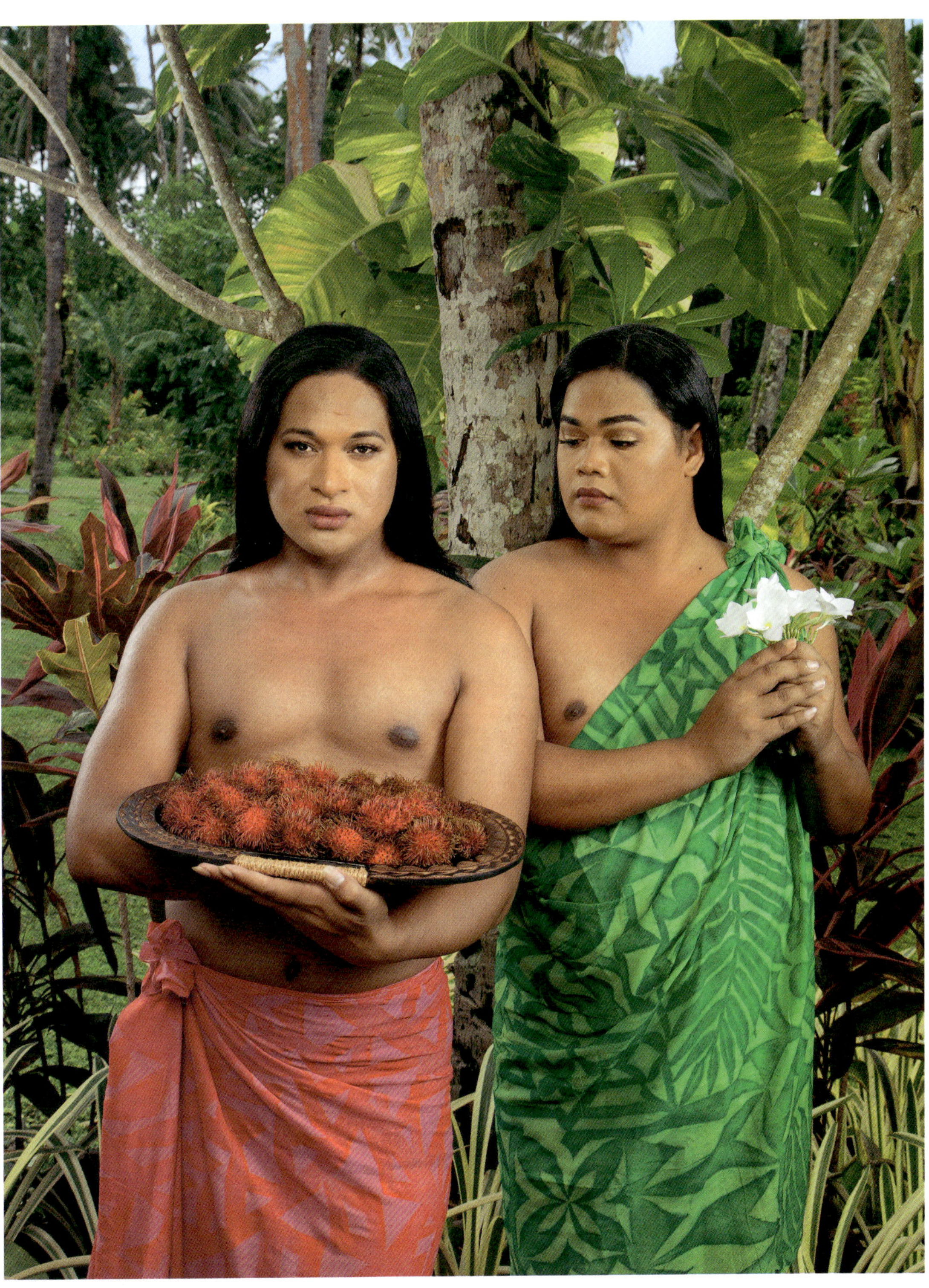

rest-and-recreation stereotypes has been a major issue for Polynesian writers, artists, scholars and politicians ever since.[2]

[TM] *Paradise Camp* has been presented in Venice, Gadigal land Sydney and Sāmoa, and each time you have created a new aspect of the project. Can you describe the variations in these past presentations?

[YK] One of the key aspects for the touring of my *Paradise Camp* exhibition is how a new work is introduced at each new location to engage local audiences. For its premier at the Aotearoa New Zealand Pavilion at the 59th Venice Biennale in 2022, the presentation was geared towards interrogating Western art history, which has become standardised throughout the world. Western art history has served as a catalyst to perpetuate and impose binary aesthetic values upon non-Western artforms, viewing them as 'primitive'. This continues to be reflected in the way objects are valued through the selection process of objects for repatriation to their rightful, non-Western owners.

The *Paradise Camp* exhibition presented at the Biennale comprised of three sections: the photographic series, which upcycles Gauguin paintings; the *Vārchive*, which highlights the backstory, including the research behind the exhibition; and a single-channel video projection entitled *Paradise Camp TV*, which showcases a variety of clips ranging from a Fa'afafine talk show and prime-time news to a *talanoa*, or conversation, between myself and the artist Paul Gauguin (where I disguised myself as Gauguin with prosthetic make-up and costume).

For the touring of the *Paradise Camp* exhibition, presented at the Powerhouse Museum in Gadigal land Sydney in 2023, I was commissioned to produce two new works, with the first being a collage series entitled *Gauguin Landscapes*, which resulted in my research into the museum's photographic collection of Charles Kerry (1857–1928). *Gauguin Landscapes* pairs the photographs of Sāmoa taken by Sydney-based photographer Kerry with the paintings by Paul Gauguin, which capture an impressionistic landscape of French Polynesia. The pairings highlight the uncanny resemblance and the seamless transition between the landscapes created from the photograph and the painting, alluding to the shared aesthetic between the two artists informed by the colonial gaze.

The second commissioned work, *BERTHA* (poetically spelt in caps to allude to BERTHA's big personality), is named after the drag alter-ego of Harold Samu (b.1961), a Sāmoan New Zealander, a gay man and a prominent drag queen who later moved to Gadigal land Sydney in the 1990s. The *BERTHA* series comprised vintage Pacific dolls that I collected from thrift stores around Tāmaki Makaurau Auckland and upcycled with new wigs and drag outfits formally worn by BERTHA to describe a chronology of BERTHA's life, where each outfit describes the year and the occasion it was worn, including the times when BERTHA appeared in Pride marches and drag shows in Aotearoa New Zealand and Australia. I'm glad that this series – which is now part of the permanent collection of the Powerhouse Museum – helped to immortalise BERTHA's legacy as a cultural icon in the Pacific.

For the touring of the *Paradise Camp* exhibition at the Saletoga Sands Resort and Spa on Upolu Island, Sāmoa in 2024, the material is presented in an outdoor exhibition and takes over the entire premises of the resort. Each work from the *Paradise Camp* photographic series is situated among the lush tropical garden as if to blend into the natural environment (fig. 9.2). There is also an emphasis in the *Vārchive*, which describes the links between Gauguin and Sāmoa and argues for how Gauguin used photographs of and from Sāmoa as foundational references for the development of his major paintings. Highlighting this relatively unknown

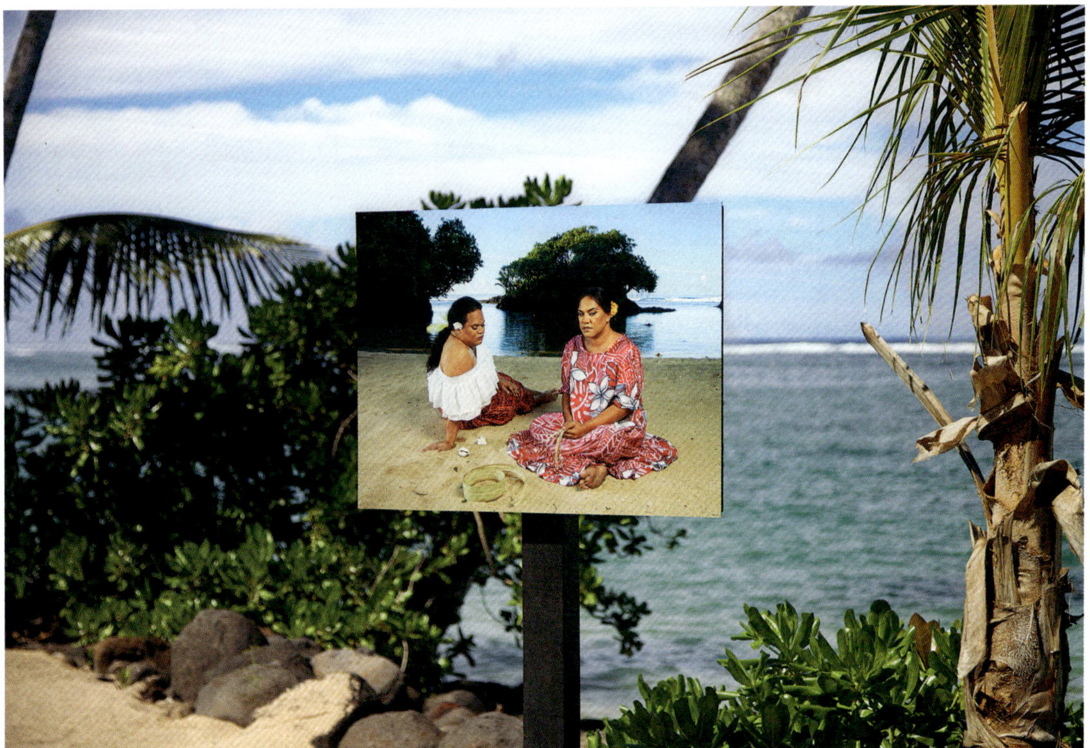

Fig. 9.2
Yuki Kihara,
Two Fa'afafine on the Beach (after Gauguin) in situ at the Saletoga Sands Resort and Spa on Upolu Island, Sāmoa, 2024

link in Sāmoa helps to bring the origin of Gauguin's paintings back to where it all began.

Each touring iteration is accompanied by a *talanoa* forum – an educational platform bringing together invited artists, curators, researchers, scholars and policy analysts to help extend the themes of *Paradise Camp* through site visits, seminars and workshops, while working closely in partnerships with various institutions of each location. *Talanoa* in Sāmoan means 'to have an open and transparent dialogue'. One of the outcomes of the *talanoa* forum was that a chapter written by each of the participants featured in a scholarly journal entitled *Lagoonscapes*, published by the Ca' Foscari University of Venice in 2023.

[TM] For the presentation at the Sainsbury Centre in the UK you have been looking into the research of Charles Darwin. How did this interest arise?

[YK] The work entitled *Darwin Drag* is built on my previous video series, *Yuki Kihara and Paul Gauguin*, featuring a range of videos of myself and myself disguised as Paul Gauguin (fig. 9.3) having a *talanoa*. We speak about a range of issues arising from the *Paradise Camp* exhibition, from addressing Gauguin's reappropriation of Sāmoan photographs to Fa'afafine and Fa'atama participation in a climate change workshop.

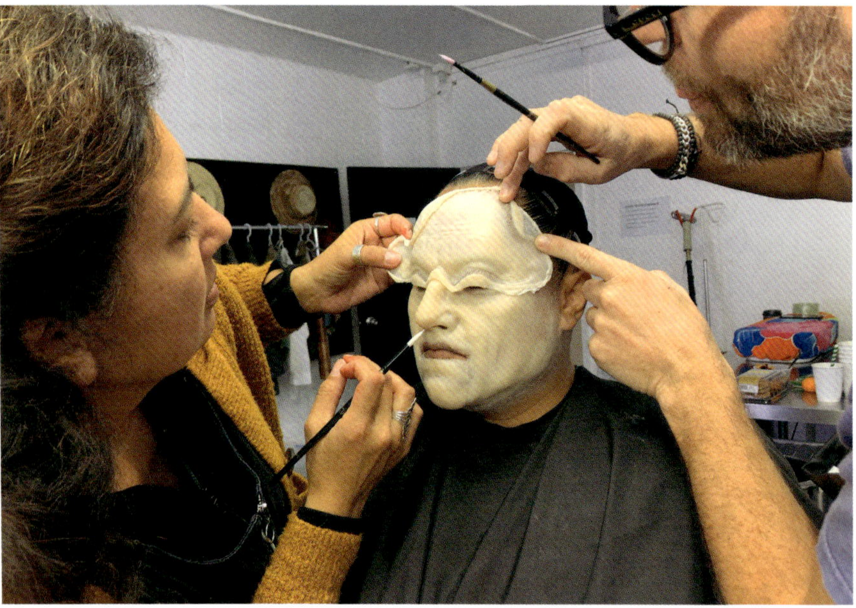

Fig. 9.3
Yuki Kihara in prosthetic make-up to transform into Paul Gauguin, 2022

The *Vārchive* presented at the Venice Biennale in 2022 included aspects of my ongoing research into sex-changing fish species found in the ocean surrounding the Sāmoan archipelago. However, I had no specific intention of evolving this research into an artwork. It was only when I was approached to tour the *Paradise Camp* exhibition to the Sainsbury Centre in Norwich, under the theme of *Can the Seas Survive Us?*, that I began thinking about the UK audience and how I might engage them through a Pacific lens as a way to bridge the gap between these two disparate places.

In *Darwin Drag*, I transform myself into Charles Darwin, heralded as the godfather of evolutionary biology. The project was born from new research revealing that the British biologist hid aspects of his research on non-heteronormative species and same-sex attraction in animals due to the prevailing conservative values of his time. I have created a humorous and campy video that features Darwin revealing to a Sāmoan Fa'afafine drag queen, BUCKWEAT, that he has been unhappy keeping his secret about queer behaviours in the animal kingdom in the closet. With the support of BUCKWEAT and friends, Darwin is transformed into a Polynesian-style drag queen and dances as one of the 'girls' in a cabaret performance where they showcase various skits featuring costumes that allude to sex-changing fish species.

Darwin Drag will be presented at the Sainsbury Centre alongside the fish specimens held at the Natural History Museum in London. Excitingly, they have specimens collected by Darwin during his time in Tahiti in 1835. One of them is a parrotfish, also called '*fuga*' in Sāmoan. The *fuga* is protogynous, meaning it has the ability to change sex from female to male. The *fuga* also spend most of their day eating algae off the reefs, which helps the corals thrive. They also snack on the hard parts of coral and their digestive process makes them excrete white sand. An excerpt from the skit about the *fuga* in the video features Darwin holding white sand with both hands and saying, 'Who doesn't love a sandy beach?'. There is currently a government ban in Sāmoa against the excessive fishing of *fuga* to help to combat the growing epidemic of algae on corals triggered as result of global warming. The survival of the *fuga*, together with many other fish species, is an integral part of maintaining coral health and a wider ecosystem in the ocean.

The aim of *Darwin Drag* is for complex theories around evolutionary biology to be presented in an accessible and creative way. I hope that my work encourages the UK audience to reconsider Charles Darwin and Paul Gauguin and how they were a product of their time.

[TM] You have spoken about how the Fa'afafine were the first to respond during a natural disaster in Sāmoa. Can you speak about how they have a caring role in society?

[YK] In Sāmoan traditional society there are four culturally recognised genders. There is Fa'afafine, meaning 'in the manner of a woman', used to describe those like me, assigned male at birth, who expressed their gender in a feminine way. I guess the closest translation to that in the Western context is a transgender woman. There are those that are Fa'atama, meaning 'in the manner of a man', who are assigned female at birth, but express gender in a masculine way. Then there is Tane, meaning a cis-gender man, and Fafine, meaning a cis-gender woman.

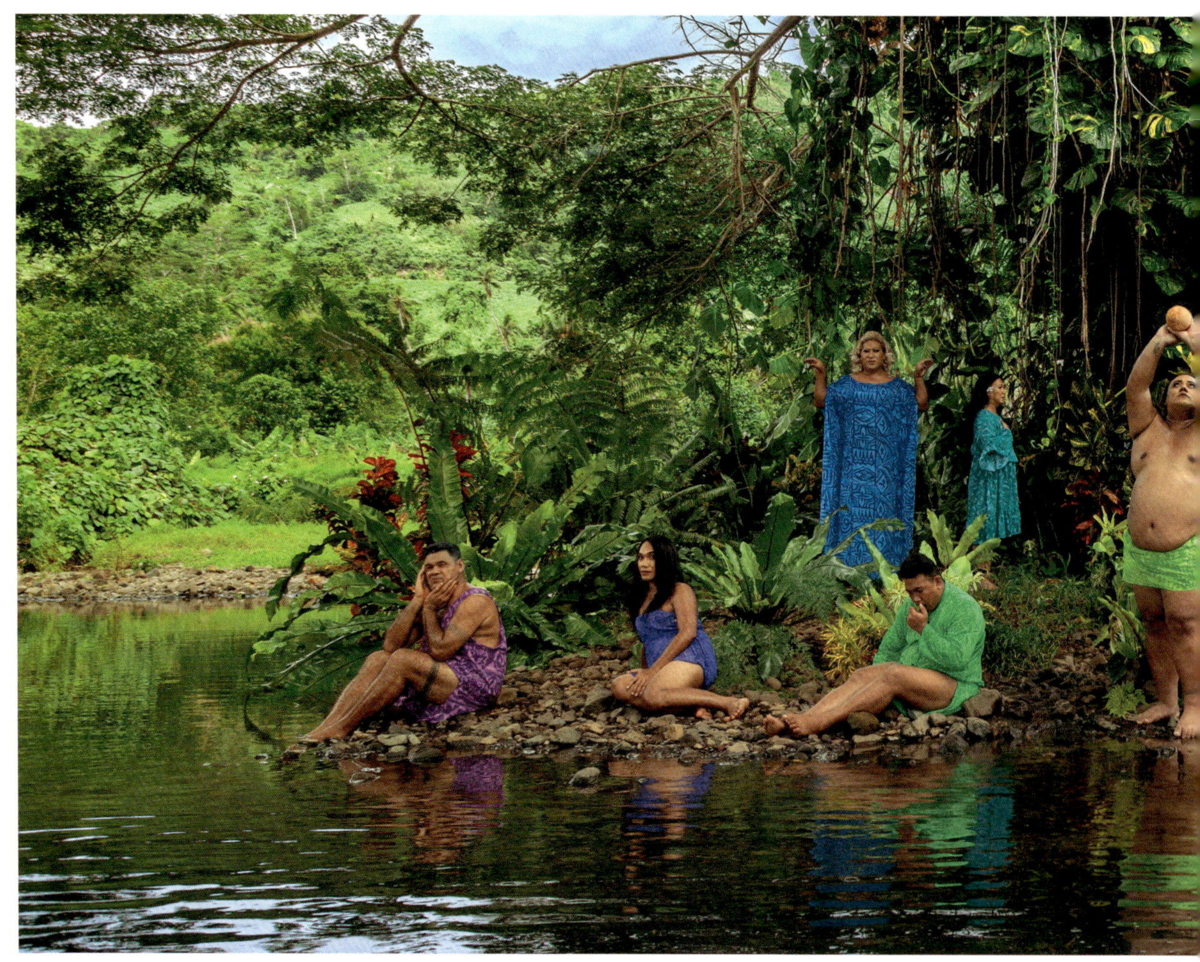

In the traditional village context, the Fa'afafine and Fa'atama are assigned the role of caregivers, and we look after young children and the elderly. Through our contribution in the community, all the four genders come together to build the resilience needed to take on any challenge. However, it was through the process of colonialism that the binary was imposed, and we, the Fa'afafine and Fa'atama became excluded. Our collective resilience was weakened and broke. As a result, the genders of Fa'afafine and Fa'atama are yet to be legally recognised. Today, the policies and legislations around climate change in Sāmoa are most often written in a binary way that excludes the Fa'afafine and Fa'atama's contribution.

[TM] Returning to *Paradise Camp*, what would you like visitors to the exhibition in the UK to take away from this piece?

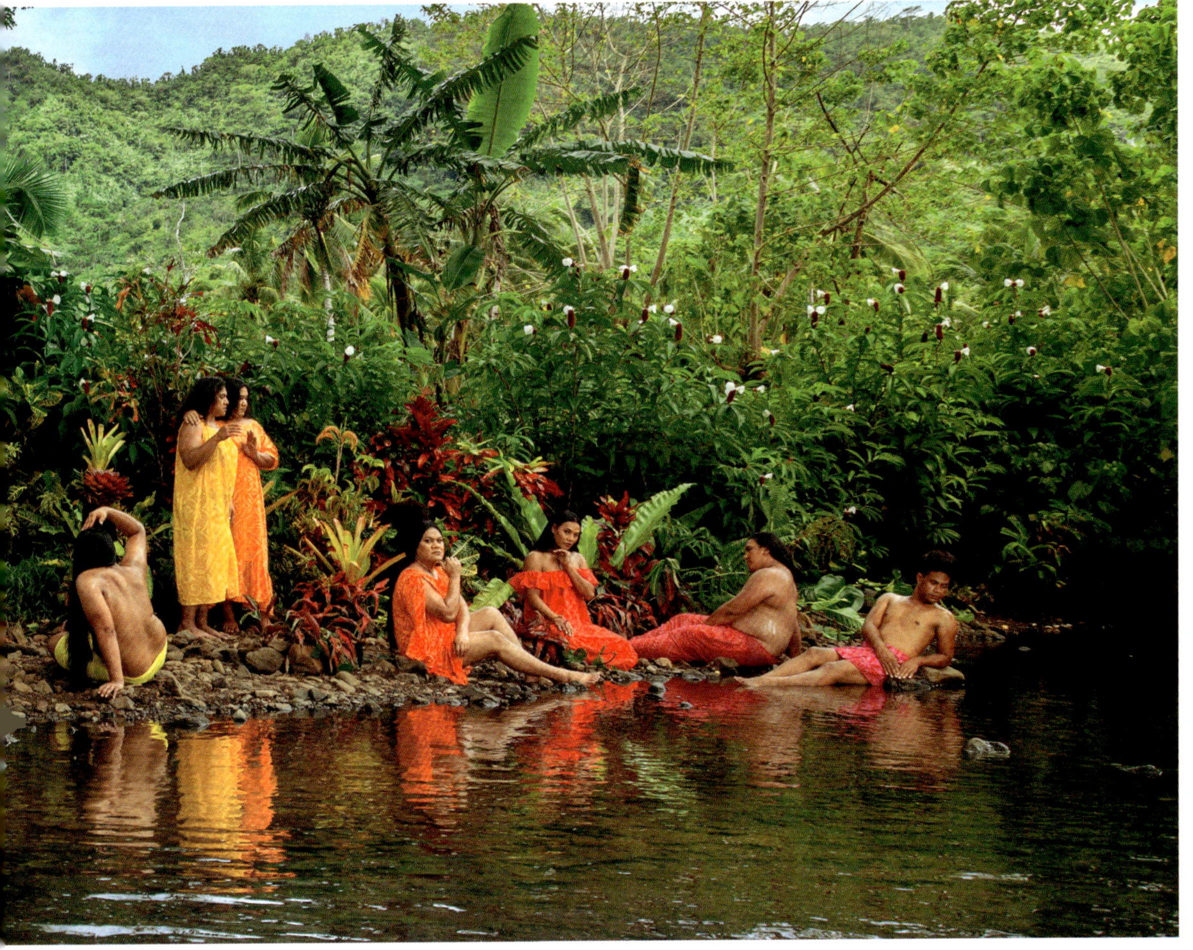

Fig. 9.4
Yuki Kihara, *Fonofono o le Nuana: Patches of the Rainbow (after Gauguin)*, 2020, C-print in four parts

[YK] In *Paradise Camp* I wanted to reclaim the fictions of Gauguin and his portrayal of the Pacific region by inserting a new narrative, by imagining a Fa'afafine and Fa'atama paradise where no one is judged for their gender or sexuality, and everyone lives in harmony with nature (fig. 9.4). As Chilean-American writer Isabel Allende once said, 'you can tell the deepest truths with the lies of fiction'. I hope that one day, *Paradise Camp* can be a reality in Sāmoa.

ESTUARINE
PATRICK FLORES

In 2024, I convened a round table on the estuary in Southeast Asia at the Philippine Pavilion of the Venice Architecture Biennale. It sought to reflect on ecological and social history through the project of the Pavilion, which centred on a particular estuary in Manila, the Estero de Tripa de Gallina, said to be Metro Manila's longest creek, around 7,600 metres long, and also one of the most clogged by improvised settlements and detritus. It runs through the cities of Manila, Makati, Pasay and Parañaque and is one of the 47 tributaries of the Pasig River, which is the main river of the capital. The estuary is connected to the Pasig through the Estero de Pandacan, Estero de Concordia and Estero de Paco (figs 10.1 and 10.2). Tripa de Gallina roughly translates to 'the intestine of a hen', thus the title of the Pavilion, *Tripa de Gallina: Guts of Estuary*.

Fig. 10.1
John Bach,
Map of Manila, 1920.
Library of Congress,
Washington, D.C.

I thought it important to generate the necessary discourse around the Pavilion from different sources. One perspective came from the curators and architects of the Pavilion. Another response was from the people living around the estuary, who serve to act as either a catalyst for change or else the cataclysm of indifference when it comes to the estuary (figs 10.3 and 10.4). A further source was the existing scholarship on the estuarine condition in the Philippines and Southeast Asia, which includes the strategy to craft, as well as the analysis of, environmental and heritage policies to cultivate a responsive and responsible public sphere.

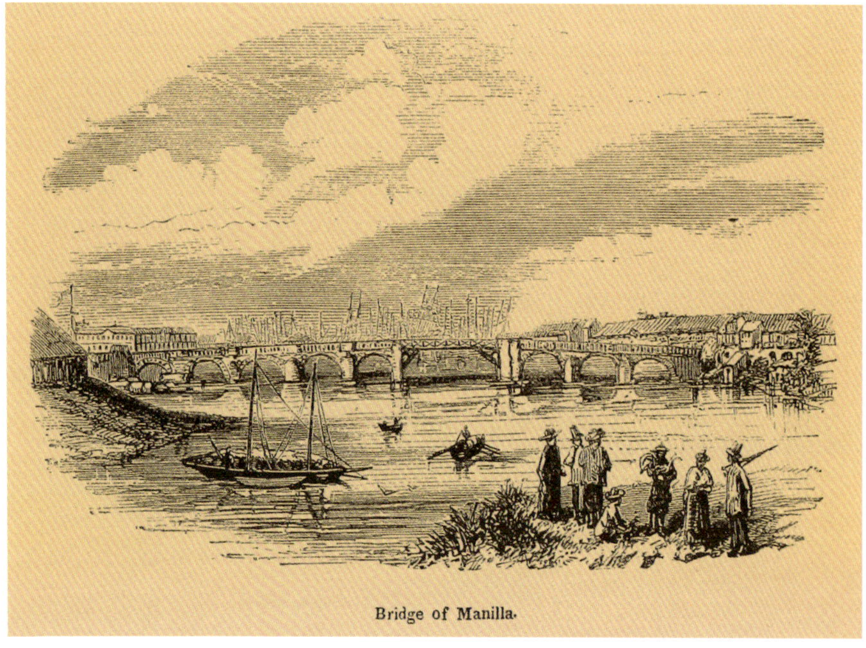

Fig. 10.2
Paul P. de La Gironière, *Illustration Showing the Bridge of Manila Across the Pasig River*, 1854.
Library of Congress,
Washington, D.C.

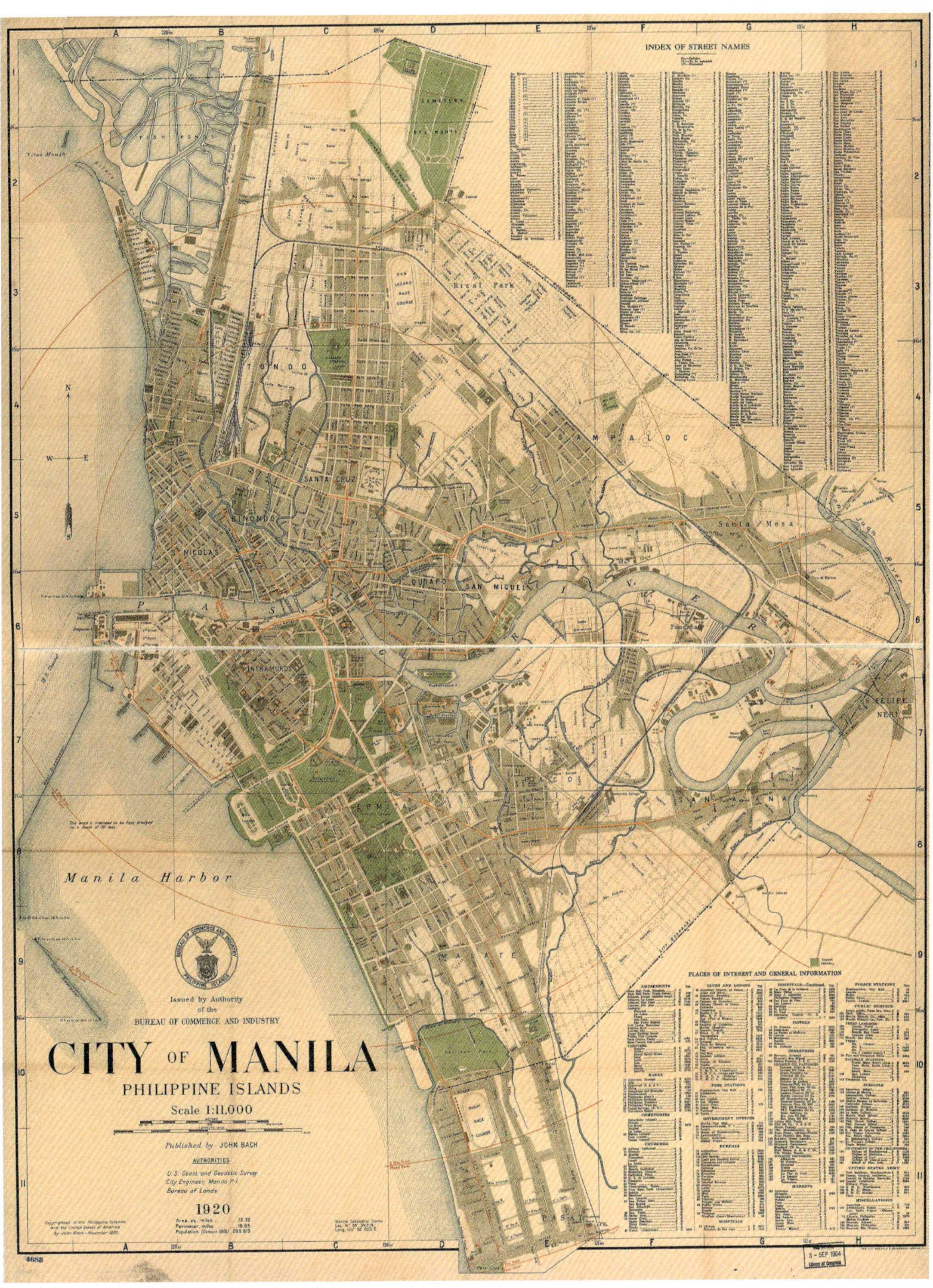

INDEX OF STREET NAMES
CEMETERY DEL NORTE
FISH PONDS
Vitas Mouth
SAN LAZARO RACE COURSE
Rizal Park
TONDO
SAMPALOC
SANTA CRUZ
BINONDO
SAN NICOLAS
Santa Mesa
San Juan River
QUIAPO
SAN MIGUEL
PASIG RIVER
INTRAMUROS
FELIPE NERI
PACO
SANTA ANA
ERMITA
Manila Harbor
MALATE
PLACES OF INTEREST AND GENERAL INFORMATION
Issued by Authority of the BUREAU OF COMMERCE AND INDUSTRY
CITY OF MANILA
PHILIPPINE ISLANDS
Scale 1:11,000
PASAY RACE COURSE
Published by JOHN BACH
AUTHORITIES
U.S. Coast and Geodetic Survey
City Engineer, Manila P.I.
Bureau of Lands
1920
PASAY

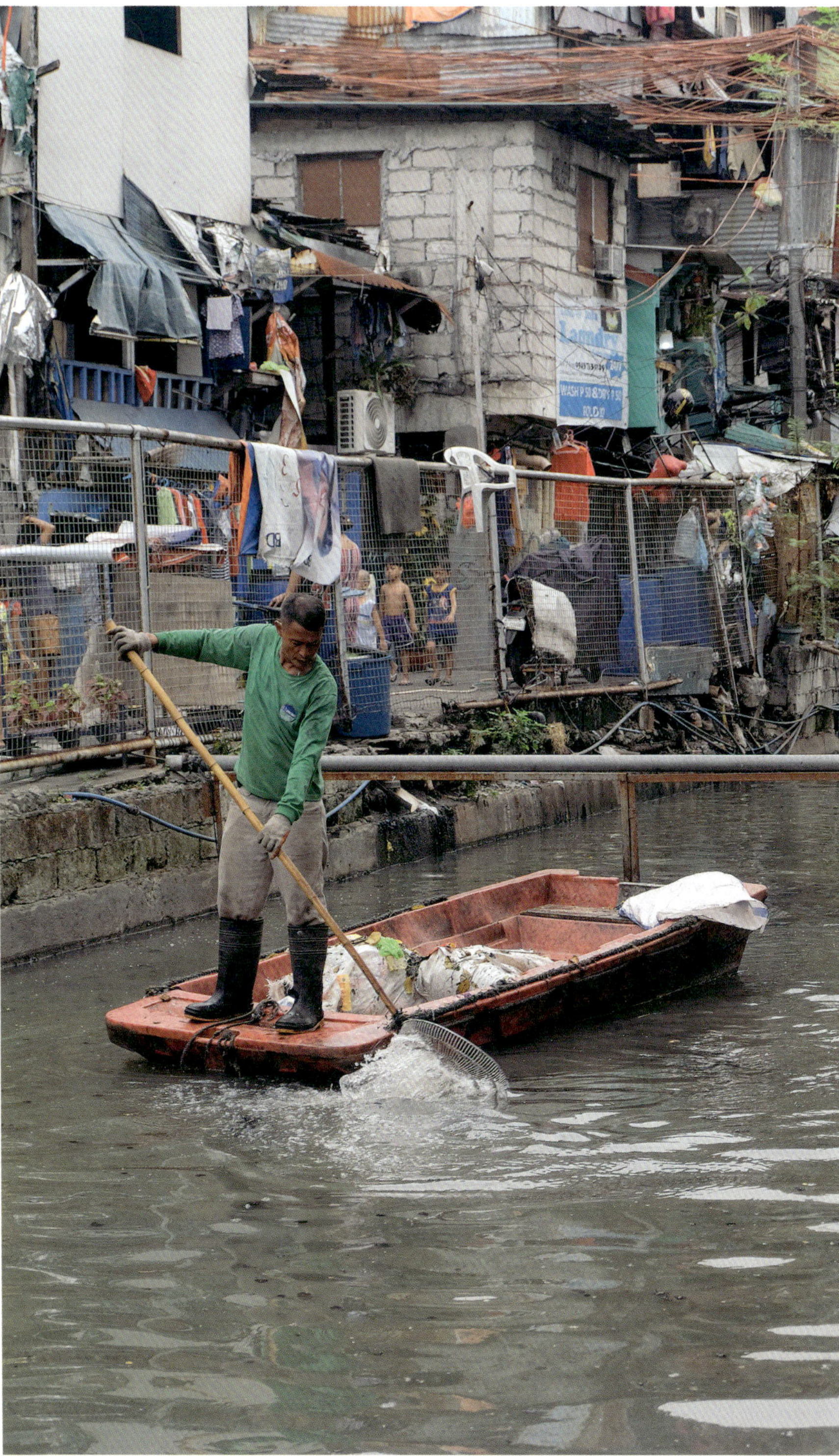

Fig. 10.3
River Warrior Arnold Pumago during his daily clean-up of Estero Tripa de Gallina along Barangay 739, March 2023

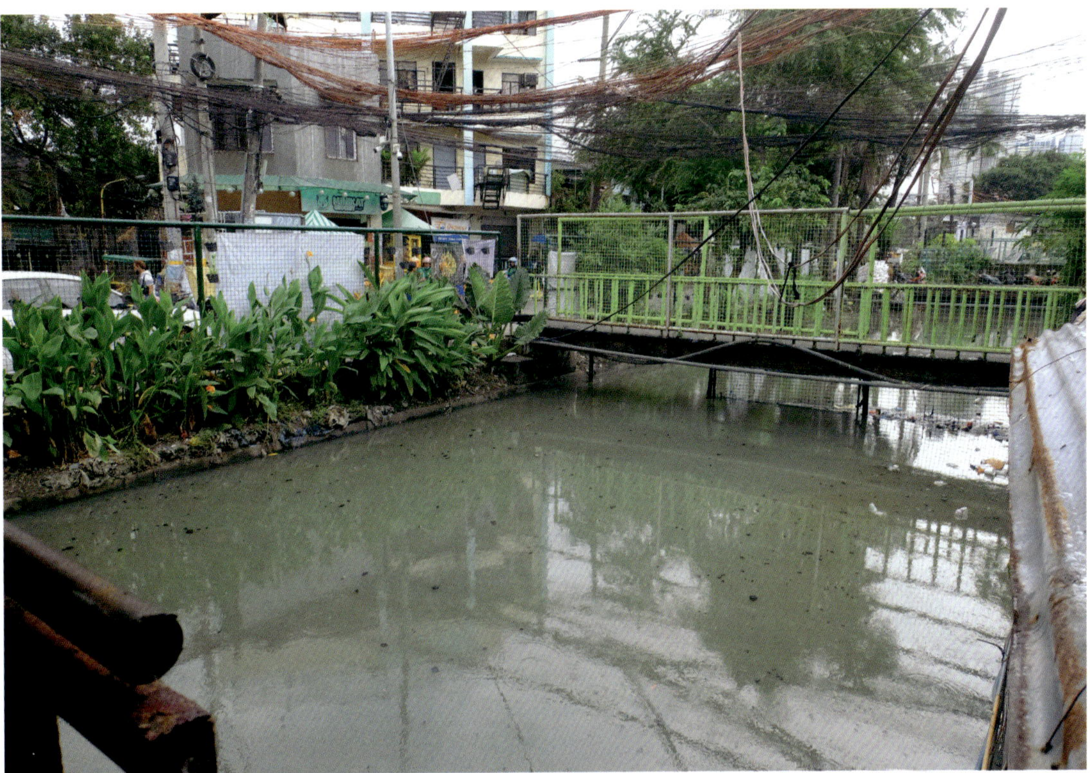

Fig. 10.4
Estero de Tripa de Gallina in Barangay 739 after the regular clean-up, March 2023

Analysis and strategy foreground the estuary as a vital node in the network of bodies exerting pressure – human beings, spirits – alongside the multiplicity of mediations – infrastructure, folklore, local government, globalisation, human and species migration. Within this context, the Pavilion becomes part of a broader historical and ecological sequence. And, at the points at which forces converge in the estuary, the Venice Biennale and interdisciplinary reflection, the previous binaries of nature and culture, local and global, architecture and society become more fluid and less preemptive. Through this round table, curators, architects, residents, scholars, development workers, urban planners and the Filipino community in Italy took part in a lively conversation on the estuary, under the aegis of a long history and a deep ecology, in the urgency of the ethnographic present and in the imagination of a possible future. It is from this matrix of relations that the Pavilion realised its presence, its bamboo structure a nexus between water and dwelling, a place of a sustainable convergence and a forum of misgivings and aspirations. In many ways, the bamboo performed the trope of the estuary as an intersectional opportunity, both a parable and a modified parabola as evoked by its form. The estuary was imagined as a geophysical entity and a motif, announcing itself through the metaphor of an intestine of a chicken.

I invited to the forum Michael Pante and Anthony Medrano, scholars of the estuary based in Manila and Singapore respectively. Studies on the estuary in Southeast Asia are nascent. Pante and Medrano in their respective projects have touched on the estuary in relation to broader contexts, such as the history of the city or ecological history. Both come from history departments and were trained as historians, but their approach is decidedly interdisciplinary. I wanted the talk to explore the potential of the estuary not only as a unique analytical rubric in the discipline of history – as differentiated from, let us say, the sea, ocean, river or lake – but also, I wished for it to strongly position the estuary as a geopoetic category, one that indexes place and yet creates its own creative energy and affective rhetoric through human activity, mythology, ecological movement and so on. While there may be variations in certain features, the estuary is typically 'an area where a freshwater river or stream meets the ocean. When freshwater and seawater combine, the water becomes brackish, or slightly salty'.[1] As a curator and art historian, I am fascinated with this instance of thresholds touching, of two bodies interspersing and causing a substance to turn into something else, its properties not settling into the rigour of object but rather becoming 'brackish' and salty though only 'slightly'.

More than straightforward conclusions and explanations, I am interested in resonances or vibrations emanating from teeming contact zones like estuaries. Through estuaries, we get a different view of historical and inventive life, and the world around and beyond us. Finally, by thinking about estuaries, we can begin to propose, what I would call, the 'civic technology' through which to motivate the community, civil servants, policymakers, urban planners, architects, designers and development workers to revive the waterways across the Philippine archipelago as vital arteries and capillaries of contemporary life. In this contribution, I try to shift the focus from the sea, which tends to be a master narrative of worlding, to one of its passageways, the estuary, to ponder the elaborate circulation of water between islands, around urban sprawls, through archipelagos. The estuary signifies the more granular aspects of the sea and complicates the rhetoric around the climate crisis with both ethnographic and aesthetic detail.

From this experience in Venice, I would like to pursue the ecology and the figurative language, or tropology, of the estuary to complicate the account of contemporaneity as mainly enabled by the ocean, the sea or the river. I am drawn to the interstitial channel of the estuary to elaborate on the theory of imbrication as a foil to the more legible politics of rupture and separation. I had earlier attempted in Venice, through the Philippine Pavilion in 2015, to speak to the state of the sea, the South China Sea in particular, in the context of the geopolitical flashpoint in which China seeks to claim a hegemonic swathe of this part of the world. Other countries around this body of water have made their claims, too, including the Philippines. The Pavilion tried to trace a longer arc of history and fathom a deeper origin of ecology and epic to underwrite, and supplement, the claim. As I look back on the effort, I realise that it was the subtle details of form such as the micropolyphony of sound, or the luxuriant texture of cloth woven around a ramshackle ship, that anticipated a moment of transcendence from the irresistible geopolitical critique.[2] This transcendence pointed to a cosmology and a contemporaneity that were both uncanny and punctual. It resonated with the title of the Philippine Pavilion for the 56th Venice Art Biennale in 2015, *Tie a String around the World*, a line taken from the 1950 Philippine film *Genghis Khan* of the singular Manuel Conde (1915–1985), a daring cinematic project on the Mongol conqueror, one of the architects of the modern and contemporary world.

I begin this preliminary discussion into the estuary with a painting and a film from the Philippines.

In the 1883 painting *Recuerdos de Antipolo* by Felix Martinez (1859–1907), we see a body of water in the foreground as a place where people gather to bathe, wash clothes and sell wares (fig. 10.5). Amid the people are a water buffalo and lush plant life. Receding into the background are houses, which indicate the

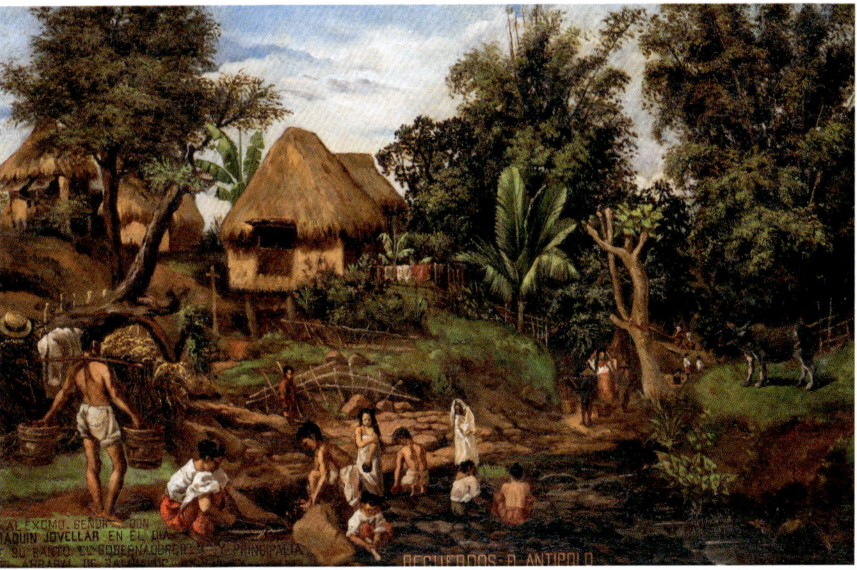

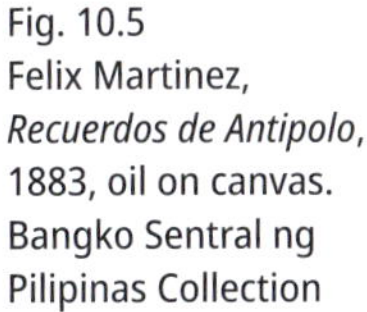

Fig. 10.5
Felix Martinez, *Recuerdos de Antipolo*, 1883, oil on canvas. Bangko Sentral ng Pilipinas Collection

existence of a village and its folk. Clothes, for instance, are hung on a wooden scaffold. Finally encompassing the foreground and the background are the sky and the clouds. As this body of water, which may well be a river, proves to be an intersection of natural forms, species and human activity, so does it play out as the edge between water and land.

The place is Antipolo, which is located outside the capital city of Metro Manila, in the province of Rizal. It lies on a plateau, or specifically on the slopes of the Sierra Madre Mountain Range. It is worth noting that the Antipolo River was earlier linked to the Marikina River, in the city of Marikina, in Metro Manila, through a creek called Bitukang Manok, translated as 'intestine of chicken'. The said creek was in turn connected to a network of creeks in the environs and has been witness to all sorts of passages, from trade to war and on to Marian processions. That the painting presents itself as a *recuerdo*, a reminiscence and a keepsake, redolent of Pierre Nora's realm of memory and Susan Stewart's longing for souvenir, leads us to regard the scene as picturesque, a miniaturisation of the empirical landscape that elicits sentimentality as well as delicacy.[3] The patron saint of Antipolo is Our Lady of Peace and Good Voyage, a black figure to whom devotees pray for serenity and safe travel. The image is said to have come from Acapulco through the fabled Manila galleon, and its fluvial procession over time has involved the Pasig River.[4] The controlling visuality of the painting pertains to ablution, or the ritual of bathing, or cleansing the body. It leads to the attentive annotation of the artwork's ties to both nature and culture and makes for a compelling ecological art history.

I would like to consider the painting alongside the film by Richard Abelardo (1902–1993), *Mutya ng Pasig*, or the *Muse of Pasig* (1950) (fig. 10.6). Central here is the Pasig River, which technically is an estuary, governed by the tides of Manila Bay and Laguna Bay. Revolving around the tale of love stolen and corrupted by desire, it privileges the river as the site of refuge and haunting. It is to the waters in the town called Matangtubig, or 'the Water's Eye', that the banished woman escapes; where her child is found by a homespun couple; where a plaintive song is heard; where the image of a ghost is gleaned. *Mutya ng Pasig*, aside from the title of the film, is also a heartfelt Philippine musical form kundiman, composed by the director's brother, Nicanor Abelardo (1893–1934); and the shape-shifting *mutya* refers to a gamut of mutations: stone, crocodile, lizard, snake, turtle, boat, nymph, fairy, goddess, among others.[5] The kundiman as a musical form is essentially about selfless, undying love that is at once personal and patriotic, a folk and art song at the same time. This particular kundiman tells the tale of a former *paraluman* (meaning 'muse' and 'magnetic needle' or 'compass'), of a

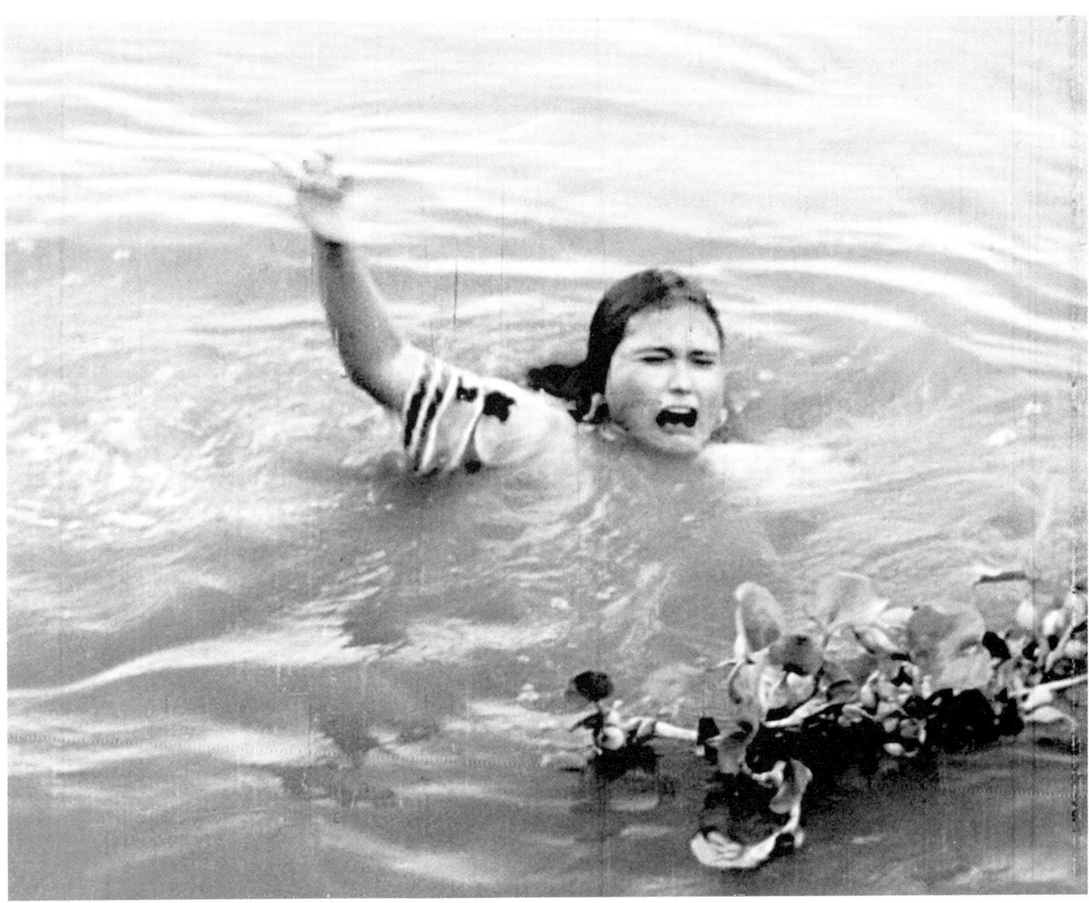

Fig. 10.6
Richard Abelardo, *Mutya Ng Pasig (Muse of Pasig)*, 1950, film still

kingdom of love, who pines for the return of life, or better still, liveliness. The director, Abelardo, was known for his contribution to special effects in Philippine cinema and worked in Hollywood in the 1920s and 1930s.[6]

The interplay of painting and film is the atmosphere that I would like to conjure as the aesthetic context of the estuary. This is a way to render the agency of the formation sensible and open to the subjectivity of drama, myth and melancholy.

The estuary in Manila is basically a system of waterways. Sometimes called estuarial creeks, they weave through an extensive city prone to natural disasters such as typhoons and floods. It is an index of a robust ecological matrix and simultaneously a complex urban crisis. As the historian Michael Pante points out:

> *Just a quick look at today's esteros reveals two crucial and intertwined postwar transformations that have caused major ecological problems. One, esteros have*

> *become the catch basin for the deluge of low-income households that have been rendered homeless by the city's unequal social structure. Two, these estuarial creeks are silent witnesses to the city's overreliance on overland mobility, which has cultivated a highly unsustainable metropolitan sprawl.* [7]

The estero, therefore, does not bask in the imagery of the idyll; it has, in fact, become the exemplary analogue of the cesspool, notwithstanding the fact that it actually is the quotidian vein of water all over the city and the archipelago.

Ecologically, Anthony Medrano deviates from the miserabilist imaginary and frames the estuary thus:

> *Estuaries are bodies of water. They are shallow and subtle, turbid and tidal.*
>
> *They are unique because their waters are the result of rivers mixing with seas – fresh mingling with salt.*
>
> *Following Engseng Ho, we might think of estuaries as 'creole waters': waters born from histories of ebb and flow, exchange and influence.*
>
> *The conditions of estuaries coproduce cultures of resilience and reliance – worlds of organisms that thrive but also delicately depend on others (and other factors) for their very existence and sometimes abundance.*[8]

Medrano fleshes out his theory of the estuary through the *buaya* (the crocodile), and the *terubuk* (the fish), in the Melaka Straits. He argues that 'like *buayas* and *terubuks*, many of the region's megacities are creatures of the estuary – hatched from the creole waters that seeded their urban origins and powered their postwar booms'.[9]

In particularising a system like the estuary, Medrano brings to our attention the lives of crocodiles and fishes within a creole ecology, inter-island and vernacular. In the same spirit, the painting of Felix Martinez and the film of Richard Abelardo not only humanise, or aestheticise, symptoms of the interventions of the Anthropocene, they also convene an interspecies assembly that may prefigure other ways of conceiving the contemporary, which is typically via oceanography, or the modern, which is condensed in terrestrial extraction and colonialism. It is rather in the contingency of tides that both modernity and contemporaneity survive. Uncannily, the song kundiman runs on a kindred contingency through the technique of *rubato*, which disrupts the expected rhythm and tempo and alternates 'accelerating and retarding for the purpose of expression'.[10] The scholar Michelle Cacho Nicolasora points out that:

> *Time, in the kundiman, is an elastic element ... the fluctuating tempo of the kundiman speaks of an emotional state that is ever changing. There is a lot of push and pull, and poignant pauses in a kundiman performance.*[11]

This essay is a critical preface to a close sensing of mingled water, which may well be the poetics of the current curatorial to be named estuarine in *tempo rubato*.

SEA INSIDE: ART AND MARINE INTERIORITY

PANDORA SYPEREK AND SARAH WADE

The sea as an unknowable and formidable entity is a cliché of the popular marine imaginary. Yet, although the oceans are undeniably vast and complex, the answer to marine biologist and conservationist Rachel Carson's famous question: 'Who has known the ocean?' is that, in fact, many have![1] Implicit in the first lines of Carson's essay 'Undersea' (1937), is that despite their mysteries, oceans contain knowledge and experiences that transcend narrow scientific understandings. Many contemporary artists, rather than succumbing to the overwhelming grandeur or the related trope of the sea as 'alien',[2] have drawn attention to the varieties of marine experience and how we may relate to these despite our own landbound 'situated knowledges'.[3] Furthermore, this 'we' becomes presumptuous when many human groups are deeply familiar with the sea, whether through boating, fishing, diving, the privileged arena of water sports or the horrors of forced or otherwise desperate maritime crossings.

A more immersive view not only challenges confounded visions of the sea by attending to its sociopolitical significance, but also enables myriad aesthetic encounters, which may not be grand or awesome, yet in their varied intimacies offer important underexplored dimensions of these vital saltwater entities and the life that dwells beneath their surfaces. The exhibition *Sea Inside*, which runs as part of the season *Can the Seas Survive Us?*, and in connection this essay, explore this alternative aesthetics to the spectacular open sea and the subaquatic sublime that are legacies of the Romantic movement, through artworks that focus on the inner spaces of the oceans, both physical and psychological.

In line with the recent academic fields of the blue humanities and critical ocean studies that have heralded a 'liquid turn' in scholarship, thinkers have examined the oceans' diverse materialities, from viscous to variegated, as models of fluidity and resistance.[4] The concept of 'tidalectics' posed by Barbadian poet Kamau Brathwaite (1930–2020) has been especially influential in rethinking art and culture through coastal rhythms,[5] while the transdisciplinary and transcontinental investigation of the 'Black Atlantic' undertaken by Paul Gilroy (b.1956) is but one formative example of scholarship turning to maritime cartographies for inquiries across histories and geographies that deploy the ocean as method.[6] With this shift away from terrestrial forms of knowledge, theorists have also reconfigured conceptions of the current ecological epoch to account for its distinctly marine and aquatic dimensions, including in relation to art and curatorial practice. For instance, Bronwyn Bailey-Charteris recently posited the 'Hydrocene' as a wet alternative to the geological premise of the Anthropocene; the entanglements of humans, communities, bodies and water in a climate-changing world leads the curator and scholar to recast Carson's question for the contemporary moment, to ask: 'Who has known the Ocean in the Hydrocene?'.[7]

Here we look beyond the oceans' immense movements, sweeping expanses and abyssal timescales towards the subtler, more intimate sites and experiences that emerge on the shores and in the depths of a seemingly uncontainable sea. These include feminist and decolonial responses to patriarchal and Eurocentric traditions of the maritime sublime, more-than-human imaginings of the sea as a site of death and regeneration, a turn to marine myth as an alternative to ecological realism and artistic expressions of the ocean as quotidian, and even mundane. Ultimately, these perspectives of the oceans lead us to question the very urge to put the sea on display, whether in the home, the museum or the gallery.

IMMERSIVE ART HISTORIES

Numerous histories have positioned the sea as an empty and horrifying wilderness in the pre-modern worldview that preceded the era of oceanographic discovery and the coinciding popular interest in seaside leisure pursuits.[8] Conversely, the following period of fascination with sea life, culminating in the 'net-haul after net-haul of strange and fantastic creatures' of the HMS *Challenger* expedition of 1872–6, becomes characterised by marine abundance.[9] And yet, whether void or bounty, these historical conceptions of the sea are both marked by a sense of psychological fathomlessness. Perhaps, then, it is fitting that the Romantic period bridging these conceptions delivered the height of oceanic awesomeness in the aesthetics of the sublime, which positioned the vast open sea as at once treacherous and intriguing. Nevertheless, this same era was equally marked by increasing popular use of the seaside, including the 'hydromania' of wild swimming by the middle and upper classes.[10] This contrast between the awe-provoking view onto the sea's surface, exemplified in the paintings of Caspar David Friedrich or J. M. W. Turner (as discussed by Courtney Traub earlier in this publication), and the physical experience of marine immersion is notable. Referring to the eighteenth-century philosophers of the sublime, Margaret Cohen and Killian Quigley explain that '[s]ublimity has difficulty underwater ... largely because the aesthetic subject, or spectator no longer has the luxury of disinterested and securely distanced contemplation which Edmund Burke and others identified as crucial to its proper functioning'.[11]

New cinematic technologies in the twentieth and twenty-first centuries arguably achieved sublime visions of the submarine zone, from the early films of Jacques Cousteau to David Attenborough's BBC television series *Blue Planet II* (2017); however, artists have pursued a noticeable visual shift from conceptions of the sea as a wondrous 'Other' awaiting exploration to foregrounding the existence of a more intimate bodily relationality. The art historical purview of the dawn of the modern environmental movement, for instance, includes artworks aligned with an ecofeminist approach, in which the interrelationships between bodies, ecologies and communities are foregrounded. This approach provided a marine counterpoint to the masculinist tendencies of the dominant interventionist land art, which the art historian and curator Lucy Lippard has called 'a kind of colonization in itself'.[12] The curator Alaina Claire Feldman observes that '[m]any artists deconstructing the ocean and environment in the 1960s and '70s were women'.[13] For instance, in 1978 Betty Beaumont (b.1946) recycled industrial coal fly ash to create an underwater sustainable fishery and marine habitat designed to benefit humans and nonhumans alike, that was not intended to be seen by human viewers.[14] Beaumont's work destabilises the ocularcentrism, and thereby anthropocentrism, of much land art and other more recent artworks installed on the ocean floor, the latter of which risk replicating the anthropocentric colonising tropes that have hitherto been tied to marine space, recalling instances such as the US Navy's Sealab projects (1964–9) and Captain Cousteau's Conshelf projects of the early 1960s, which attempted to establish underwater habitats to allow divers to live and work inside the deep sea.[15]

Critical theorists have recognised the inseparability of bodies and oceans, from Teresia Teaiwa and Epeli Hau'ofa's assertion of Pacific Islander sovereignty summed up in Teaiwa's statement, 'We sweat and cry salt water, so we know that the ocean is really in our blood',[16] to hydrofeminist Astrida Neimanis's and others' insistence that our continuity with other bodies of water disrupts the 'dry, if convenient, myth' of discrete individualism.[17] Acknowledging this permeability, the *Silueta Series* (started in 1973) of Ana Mendieta (1948–1985), in which the artist used her own body to variously interact with the earth, features a photograph documenting a sunken imprint of the artist's body in the sand that is filled with red pigment slowly draining out into the sea in a way that symbolically conflates bodily fluids with seawater and the tides (fig. 11.1).

Likewise situated on the shore, *Bodyshells* (1972) by Heidi Bucher (1926–1993), presented in *Sea Inside,* comprises a film of pearlescent sculptural forms that ensconce the bodies of dancers (in fact, Bucher's own family members) as they move playfully along Venice Beach to the sound of lapping waves, both

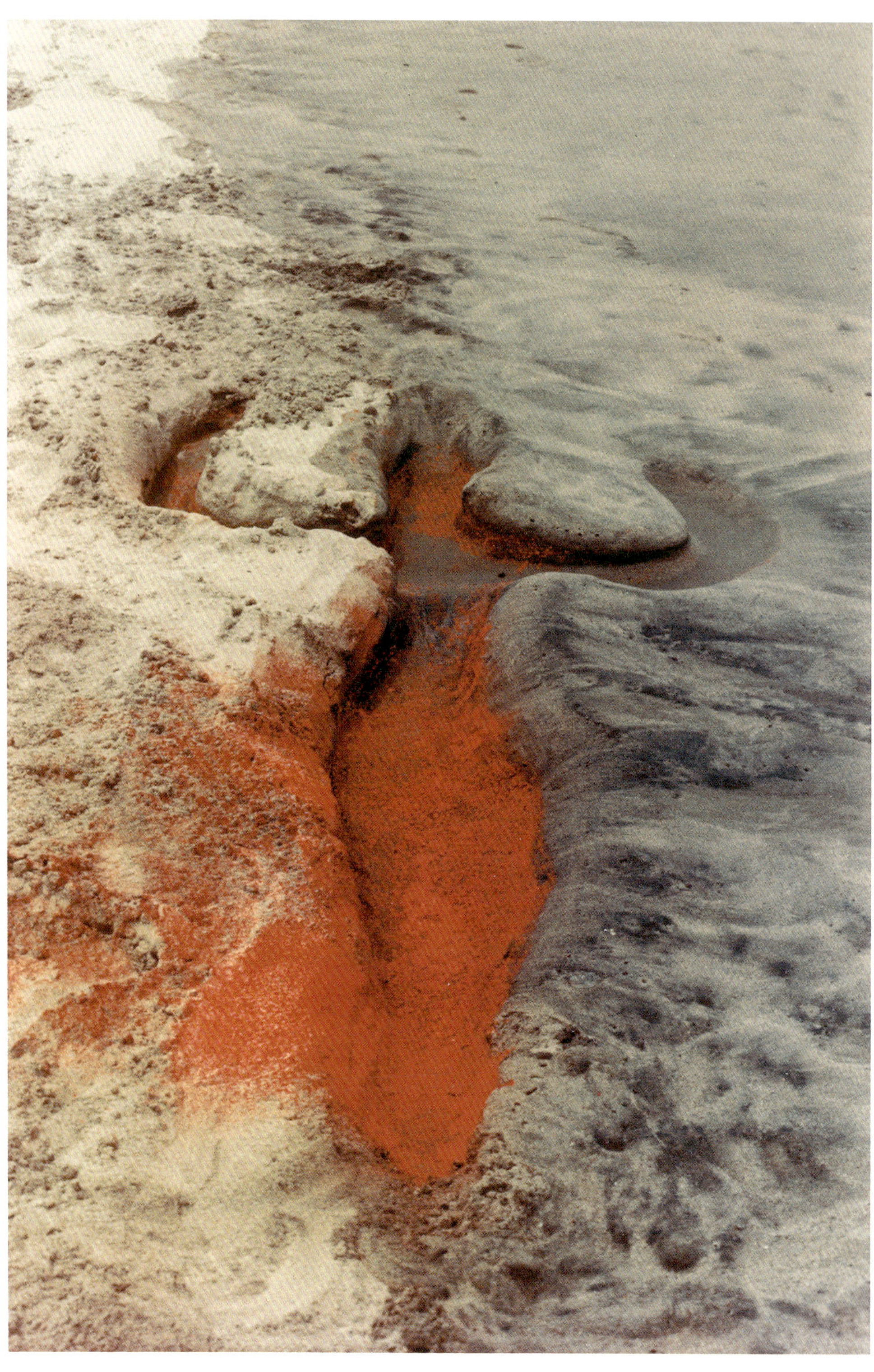

Fig. 11.1
Ana Mendieta,
Untitled (Silueta Series),
1976, photograph.
The Estate of Ana
Mendieta Collection

protecting and connecting these abstract feminised bodies to the ocean. In contrast with the landscape engineering effected by many earthworks, Bucher's wearables reference fashion garments and their proximity to the body.

Contemporary artists have continued to explore the relations between bodies and water to transcend notions of their separation, as in Evan Ifekoya's (b.1988) mindful immersion in the secluded rocky pools of Iceland, portrayed in the video *Contoured Thoughts* (2019; fig. 11.2). Here immersion is physical and bodily as much as psychological and spiritual, as an activation of the artist's dual identity as a shamanic practitioner named Oceanic Sage. In this respect, the artist at once holds tools for soul retrieval and shadow work and is themself held by the medicinal sulphurous springs.[18] As such, the duality of this otherwise unpeopled northern landscape and connections to Ifekoya's own Yoruba heritage debunks the aesthetic detachment from the land fundamental to the experience of the Romantic sublime. Rejecting the separation of 'nature' and humanity, artists and thinkers have recently foregrounded such 'entanglement', which, in the

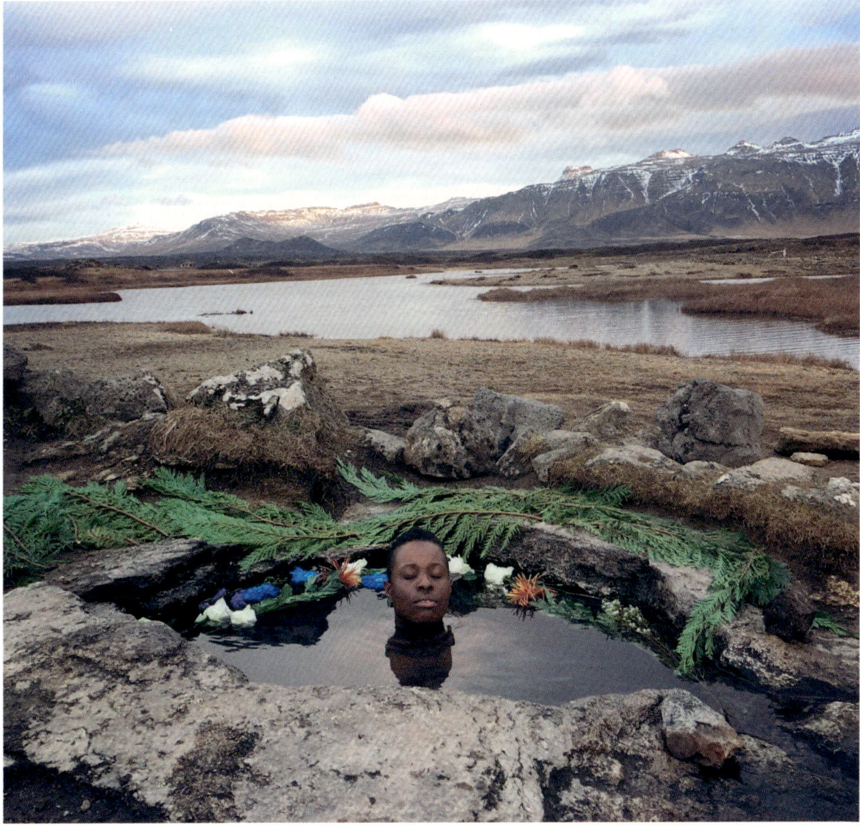

Fig. 11.2
Evan Ifekoya,
Contoured Thoughts,
2019, film still

wake of the oceanic turn in academia, Melody Jue and Rafico Ruiz suggest be reconfigured as *saturation*, to counter 'the terrestrial bias of contemporary theory' through the adoption of a more fluid and liquid model to think (and make) with.[19]

FROM WATERY WOMB TO UNDERSEA TOMB

In line with the patriarchal European tradition of marine otherness, the sea has often been characterised as feminine and maternal, from the nineteenth-century historian Jules Michelet equating seawater with mother's milk to human uteri being envisioned as mini-oceans.[20] However, the generativeness of the sea goes far beyond such anthropomorphism. Despite our briny beginnings, Neimanis maintains that 'human reprosexual wombs are but one expression of a more general aqueous facilitative capacity'.[21] Correspondingly, artists have explored expanded gestations in relation to the oceans and marine life. In a recent body of work, Laure Prouvost (b.1978) has explored maternal entanglements with the sea, frequently centring on the octopus as a surprisingly relatable mother figure. In most octopus species the female starves herself and dies after brooding her first and only set of eggs.[22] Prouvost's sculptures see the human-cephalopod separation deteriorate, with octopus arms dripping with pendulous breasts instead of suckers or holding cups and bottles at the ready for nourishment. The artist's series of 'Cooling Systems' abstracts these connections and makes plain their broader ecological import, with both mammaries and tentacles acting as standalone symbols of necessary nurture in the face of planetary crisis. Made from Murano glass, these delicate fountain-sculptures become surreally speculative devices, complete with a small watercolour illustrating their use. *Cooling System 6 (for global warming)* (2018; fig. 11.3) features eight pink glass breasts emerging out of an ultramarine base in an octopoidal roundabout formation echoing traditional decorative arts, such as the exquisite jellyfish chandeliers by Art Nouveau sculptor Constant Roux (1869–1942); in the accompanying diagram the ruby-red nipples spout essential liquid over a desperate human inhabitant of a rapidly heating climate-changing world.

Prouvost's work amusingly gestures towards the material and ecological inseparability of life that equates human gestation with the oceans: just as the symbiotic relationship of mother and baby defies the modern ideal of the individual, human life relies on the sea and therefore it is crucial that we equally tend to its needs.[23] In Kasia Molga's *How to Make an Ocean* (2019–present), the artist and designer does just that, but on a microcosmic scale, experimenting with growing marine species – namely, North Sea algae – in a series of tiny glass vials. Instead of amniotic fluid, the surrogate medium for seawater in these mini-oceans is the artist's own tears. Hence, new life emerges out of devastation, in this case tied to the artist's own grief, both personal and arising out of climate anxiety.[24] Molga's project acknowledges the mediation of both the environment and her emotions by technology and the media; in response she designs a 'moirologist bot' – an AI version of a professional mourner – programmed to mimic this ancient figure by eliciting emotional responses through algorithmically selected video footage.

The interface between technology and so-called natural gestation is likewise foregrounded in *Ghosting* (2023) by El Morgan (b.1978), a video that shows footage of baby jellyfish developing under a microscope, set against a soundtrack of the artist's telephone call with an IVF storage facility. The interaction, in which Morgan's earnest query into her frozen embryos and 'how they are doing' is met by a series of transfers to different departments before getting through to anyone who can help, presents an absurd counterpoint to the life-changing possibilities afforded by fertility treatment, linking these 'miracles of creation' to a Kafkaesque bureaucracy. Such transhuman alignments, whether

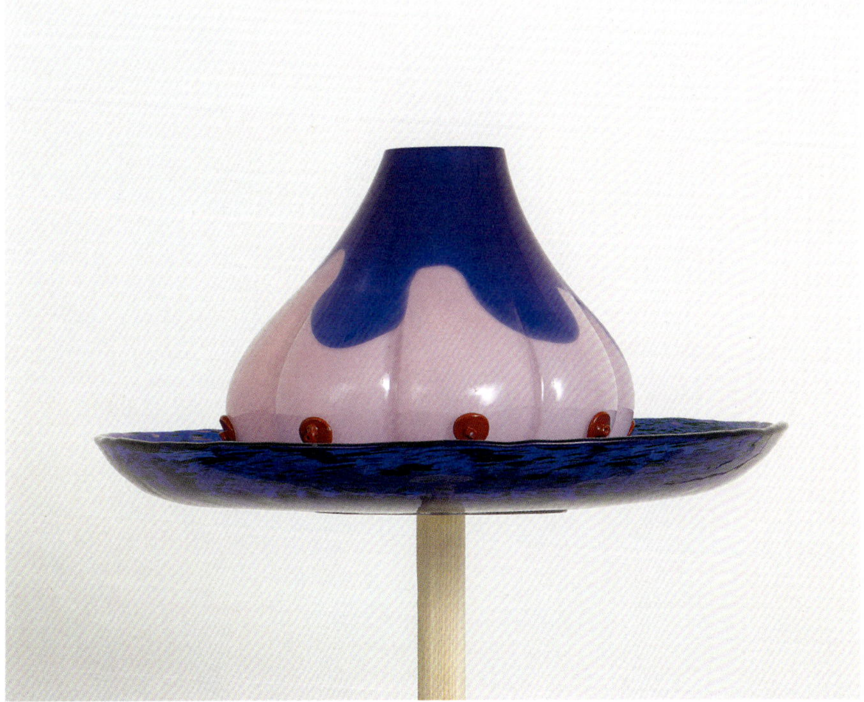

Fig. 11.3
Laure Prouvost, *Cooling System 6 (for global warming)*, 2018, glass, framed drawing and wood stick with watercolour (detail)

through technological or nonhuman overlap and collaboration, amount to a type of queering, blurring the boundaries of the human and corresponding gendered roles.

All of these works signal the unquestionably fine line between life and death, and yet in relation to the oceans this threshold is especially fluid. In the watercolour *Oceanic Feeling* (2024) by Chioma Ebinama (b.1988), a larger-than-life-sized feminine figure becomes a habitat for various sea life, real and imagined: her submerged body, gently modulated by layers of watercolour, is surrounded by squids and mermaids, her limbs navigated by tiny skeletons riding fierce-looking fish. Mounted horizontally on a coffin-like concrete plinth, the work suggests an ocean grave, and yet a very lively one, evoking the whale fall, in which the cetacean's carcass provides vital nutrition for an entire ecosystem of bottom dwellers, and hence figures the sea as a site of multispecies death as well as new life.

Victor Ehikhamenor (b.1970) addresses the history of the transatlantic slave trade via the slave ship hold, a site of unspeakable horrors and a final resting place for many. The installation *Do This in Memory of Us* (2019–20) features a giant tapestry sewn from over 10,500 plastic rosaries, in hues of cobalt, turquoise, black and white, depicting the plan of the hold overloaded with human cargo. This beautiful object with its distressing imagery appears through its inverse on a mirrored floor spotlit in a darkened room. Stepping onto the flexible mirror, which yields underfoot as if walking on water, the viewer looks down to see themself bearing witness to the scene. A score of 'Amazing Grace', sung by the Benin City Choir in the local language of Edo, further contributes to the immersive experience of the piece, and its complicated mixture of trauma and spirit to emerge out of colonialism's legacies of violence and resistance.

MYTHIC ECOLOGIES

It is notable that the shameful history of slavery has fostered a plethora of artworks responding to the related myth of Drexciya. Devised by the eponymous Detroit techno group (1992–2002), this Afrofuturist legend of a Black Atlantis populated by the descendants of enslaved pregnant women who were thrown or jumped overboard in the course of the Middle Passage has provided inspiration for numerous visual artists. Albeit influenced by the worldbuilding potential of this luminous vision arising out of horrendous cruelty, a younger generation of artists is increasingly turning to other mythical beings and their capacity for exploring issues of identity and being in the world in relation to traumatic histories. In Ebinama's poetic watercolours featuring mermaids and sea nymphs, the artist envisions these hybrid figures as relating to queerness and neurodiversity, with the return to the sea equivalent to a sort of 'unmasking', a deliverance from the societal pressures to suppress difference or heterogeneity (fig. 11.4).[25] She writes, 'the hybridity (half-fish, half-woman) is an interesting (perhaps ancient) means of describing ways of being that do not fit into simple categories'.[26] This type of re-envisioning of such creatures offers new ways of questioning narratives of normativity of body and mind, and their place in relation to the environment. In her book *Disabled Ecologies* (2024), Sunaura Taylor examines both the disabling impact of ecological contamination on human and more-than-human life, as a close corollary to environmental racism, and, conversely, the ablism implicit in representing disability as a 'cautionary tale' or metaphor, for example for 'diseased' ecosystems.[27] These perspectives are especially relevant in relation to the complex webs of life within the oceans and the ecological damage affecting the sea, testifying to the importance of thinking otherwise for addressing these wicked problems.[28]

Relatedly, Gabriella Hirst (b.1990) has interrogated the myths of men being swallowed by whales via a series of mixed-media works. Cumulatively titled *Ambergris* (2023–present), the project highlights the exploitation of these cetaceans as a precursor to the petrochemical industry. Reimagining the nineteenth-century whalers' craft of scrimshaw for the twenty-first century,

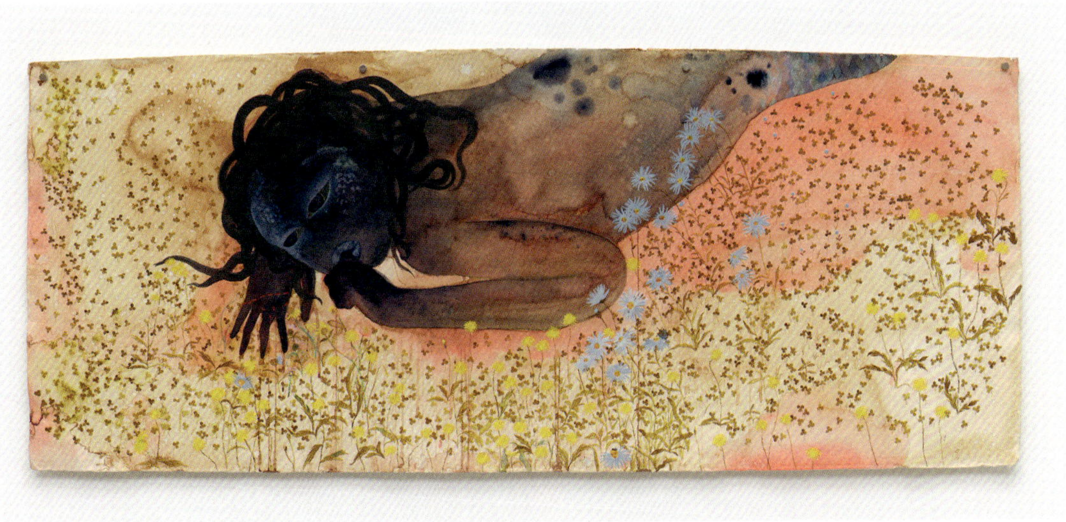

Fig. 11.4
Chioma Ebinama, *a hụrụ m gị n'anya (i see you in my eye)*, 2023, watercolour, sumi ink, gouache and coffee on handmade cotton rag paper

Hirst delicately carves images of marine species as found in life and as depicted in popular culture into polyurethane sculptures that reference the domestic and industrial products emerging from the bodies of these marine mammals, exploited for human benefit throughout history. Hirst links the desire behind this outward pursuit of capital to the erotics of being inside the body of the whale, explored variously through the artist's footage of museum visitors walking in and out of the hinged jaws of the preserved body of the 'Malm' whale in the Gothenburg Museum of Natural History, Sweden (in which a couple is said to have once been found having sex) and the genre of internet memes showing shirtless muscular men comically emerging from whale blowholes, which have formed part of Hirst's wide-ranging artistic research into this phenomenon.

Hirst's project is notable for bringing far-reaching tales, from the Book of Jonah to the Disney animation *Pinocchio*, into dialogue with environmental and industrial histories and their accompanying cultures. As such, it and related work offer a counterpoint to the focus on documentary realism within the burgeoning field of ecologically orientated art and curation, and thereby meet calls by Neimanis and colleagues to 'address problems of compartmentalization, alienation, and an unhelpful focus on technocratic management rather than values and ethics'.[29] Correspondingly, the series of sea shanties by the artist Harun Morrison (b.1981; reproduced in this volume) reworks the traditional form to unite the ballad's associations of sailorly camaraderie and nostalgic yearning for lives and loves left ashore with consideration for the experience of other forms of sea life, whether biotic, as in the case of whales slaughtered in the Faroese *grindadráp*, or mythological figures of selkies and mermen. Morrison conceives of these songs as themselves vessels, which carry multiple coinciding motivations, emotions and memories.[30] By containing mythical and environmental issues alongside these themes, these shanties become increasingly capacious, embracing ancient and nonhuman positionalities on a continuum with both traditional and contemporary concerns.

Morrison's work imbues current concerns for marine biodiversity with the lessons of antiquity. Both selkies – creatures that shapeshift from seal to human – and merfolk – fish-human hybrids – frequently serve as warnings against crossing the terrestrial domain into the aqueous, and hence the tragedies that befall those who pursue them function as potential harbingers of humans' hubristic exploitation of the oceans.[31] At the same time, they can be seen as restoring an ecological sensibility to myths, which may have become detached from their original contexts within retellings.[32]

THE QUOTIDIAN OCEAN

Such folkloric incursions into seafaring life speak to an imaginative dimension that is not limited to the sphere of the sea itself but can also transpire in locations and relations that are closer to home. For instance, the coloured pencil drawings of Shuvinai Ashoona (b.1961) show everyday life in the Canadian Arctic enmeshed with Inuit mythology, such that even a routine dental check-up becomes wondrous, as in the artist's *Composition (At the Dentist)* (2022). However, as the writer and artist Tarralik Duffy observes, Ashoona's 'drawings cannot simply be described as imaginary worlds intermixed with reality. ... Ancient Inuit beliefs are replete with Unusual beings such as those seen in Shuvinai's work', so that we might be 'left to wonder whether these curious creatures are merely fantastical or perhaps something more rooted in reality than we dare acknowledge'.[33] Another example is Ashoona's *Composition (Octopus Sedna Transformation)* (2023), which depicts the Inuit goddess of the sea and marine animals as crossing the gender binary as well as divisions between sea and land, legend and daily life (fig. 11.5). Permeations of the extraordinary into the ordinary and vice versa are therefore a notable theme emerging in wide-ranging contemporary artistic practices.

Ashoona's work demonstrates how for some Indigenous communities the sea is integral to daily life and culture. In addition to portraying the wondrous in everyday scenarios, other artists locate magical properties within the objects of the sea, brought inside for various purposes. A series of sculptures crafted from abalone shells by Tyler Eash (b.1988) at once speaks to matrilineal descent – these 'grandmother shells' are passed on from female elders in West Coast cultures including Eash's own Maidu heritage – and connects to vernacular culture, such as jewellery – the artist at once carves earrings from the shells and turns them into surrogate body parts and prostheses such as masks. Small enough to hold in one's hand, these traditional items of trade with their stunning iridescent interiors shapeshift across scales and species; embodying the eyes of ancestors and the stars in the sky, they become a cosmos unto themselves.[34]

The awe associated with maritime sublimity may haunt the history of Western art, but recently numerous contemporary artists have sought to tame and domesticate the sea, recontextualising it within the familiar and familial and reimagining it on a human scale in playful and poetic ways.[35] Such work often uses humour to highlight the absurdity of these endeavours. Characterisations of the oceans as being filled with marvellous lifeforms that are hard to comprehend have been subverted by artists through performance work that seeks to bring

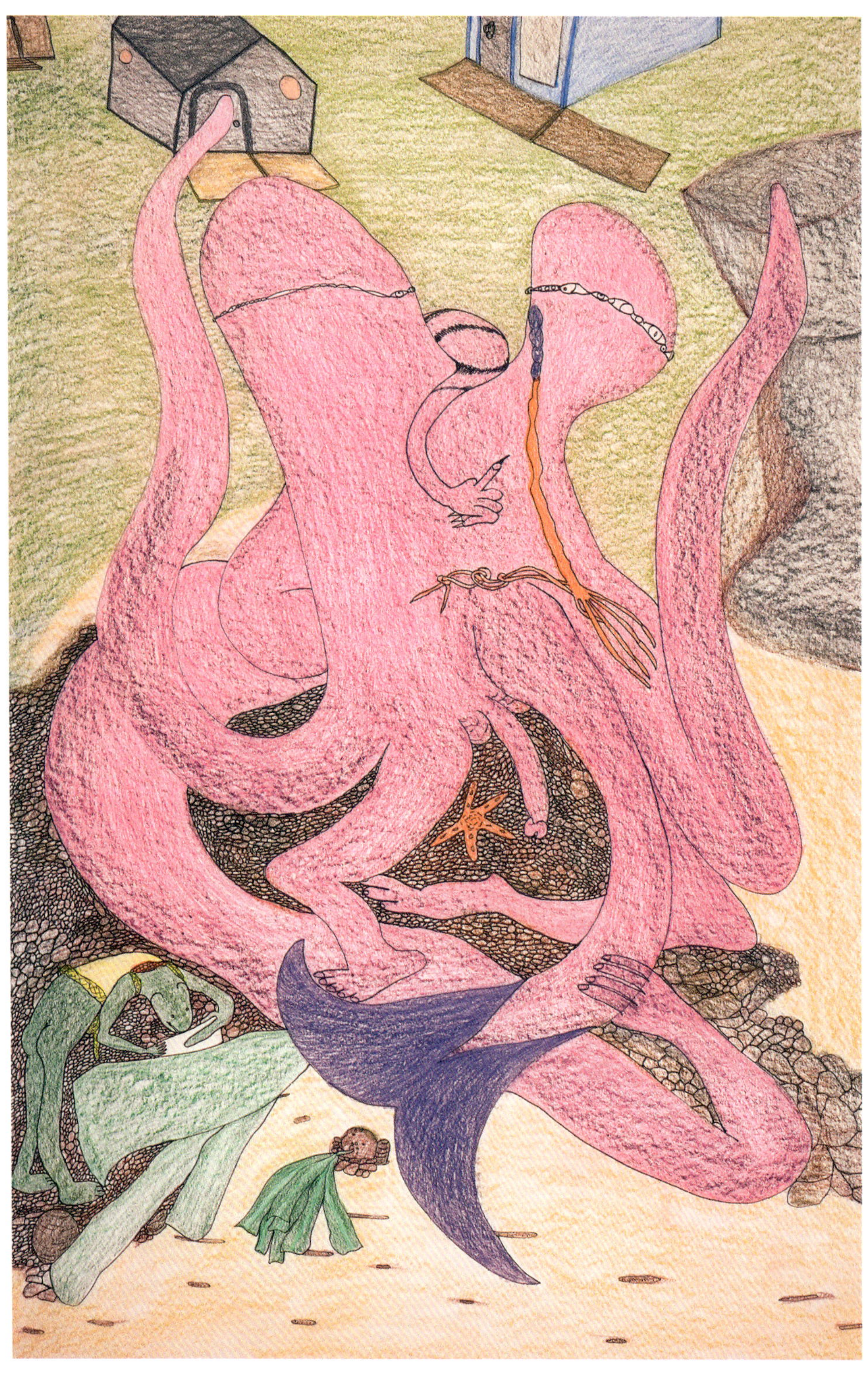

Fig. 11.5
Shuvinai Ashoona, *Composition (Octopus Sedna Transformation)*, 2023, coloured pencil and ink on paper. Sainsbury Centre

sea creatures into closer relation with human realms of experience, foster affinity and even promote wildlife protection. As such, many artists have cast the sea in relation to banality rather than myth and fantasy, whilst challenging the human-centredness intrinsic to ocean exploitation.

In *Humpback Whale* (2016), Marcus Coates (b.1968) performs whale calls from the confines of the half-filled bathtub of a family home, rubber bath toys and shower gel neatly cropped in shot. This act of imitation can be read as an attempt to bring a marine animal closer to the artist's daily domestic experiences, or even to better understand what it might be like to be a whale, in turn reducing any sense of perceived distance or difference. Yet the incongruity of imagining a massive, majestic marine mammal through such modest and makeshift means is as charming as it is humorous, suggesting an alternative agenda that at once pokes fun at anthropocentrism and highlights the restricted limits of human understanding. Coates's work additionally gestures to an alternative method for the creation of knowledge about sea life that emerges through imagination and embodied performance. As scholar Ron Broglio has observed in relation to Coates's wider *oeuvre*, 'In loosing the tethers of what it means to be human, we find new avenues and lines of flight by which to traverse the un-thought of thought', offering creative possibilities for conceiving and enacting human-marine life relations.[36]

Skye Turner's multimedia forays into the lifecycle of 'Lady Sockeye' the salmon explore the pitfalls and possibilities of anthropomorphism. The artist dressed up as a salmon to raise awareness about the anthropogenic disruptions to these creatures' lifecycles and is shown in a video nervously entering a suburban home in an attempt to find a suitable spot to spawn. The work mobilises a form of 'animal drag', most recently described by the artist-academic Nicola McCartney as 'an activist performance in which a human critically and consciously performs "animality" in order to de-centre humanism, with dress and adornment as key aspects of its dissidence'.[37] Turner's humorous performance in a cumbersome, comic costume thereby has earnest intent, embodying what the ecocritic Timothy Morton has called a 'playful seriousness' to show how attempts to decentre humans and better relate to wildlife are necessary, especially in ecologically troubled times, but are nevertheless always somehow flawed.[38]

In another effort to challenge the presumed alterity of marine animals, the artist Shimabuku tried to imagine his way inside the minds of octopuses to learn more about their favourite colours and shapes. Putting the results of his cephalopodic experiments on display in a manner recalling a museum vitrine,

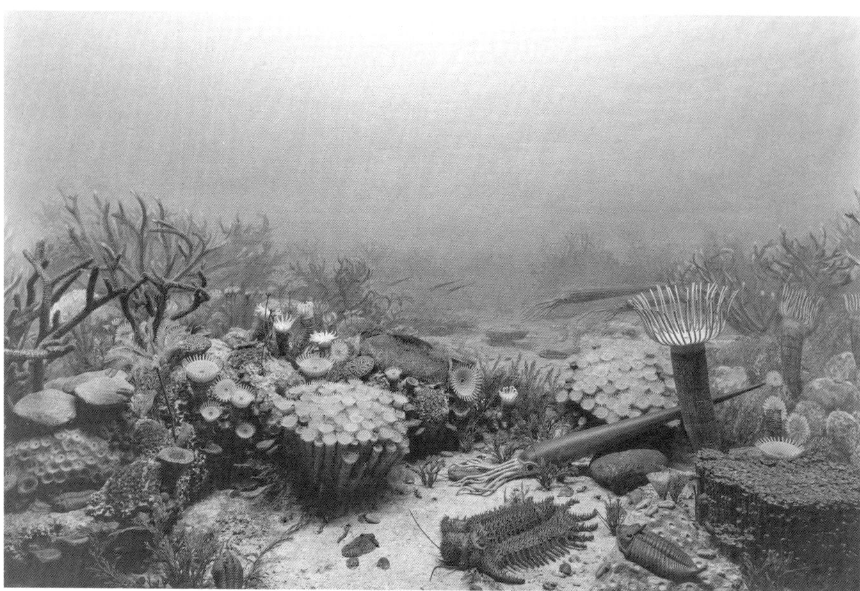

Fig. 11.6
Hiroshi Sugimoto, *Devonian Period*, 1992, from the *Origins of Love Portfolio*, 2004, gelatin silver print. Sainsbury Centre

Sculpture for Octopuses: Exploring for Their Favorite Colors (2010) suggests that octopuses each have their own taste and aesthetic judgment, positing that these nonhumans also possess inner minds. Shimabuku's work can be understood as a collaboration between artist and octopus, with the resulting installation becoming what the art historian Jessica Ullrich has called 'interspecies art'.[39]

THE SEA ON DISPLAY, THE SEA INSIDE

Shimabuku's sculpture indicates the important role of display in the attempt to bridge human and marine worlds. When it comes to taming and containing the sea, the aquarium presents the domestication of the ocean par excellence. As a Victorian parlour-room curiosity that facilitated viewing the wonder of marine life from the safety and comfort of an armchair at home – or amongst society in the public hall – it put ocean life on display and literally brought the sea inside. The aquarium, conceived by the art historian Marion Endt-Jones as 'a moving work of art',[40] conflated spectacle and proximity, captivity and wildness, becoming an alluring display model that has been influential to artists and curators since its inception as a locus of critique and contemplation.[41] This display method highlights a paradox also shared by museums of oceanography, maritime history, whaling, fishing and natural history in that to exhibit the sea is to bring it indoors to the dry air of the gallery. While these endeavours seek to enhance knowledge and familiarity, such a disjuncture can ironically reinforce distance and separation, affirming allusions of mastery and the prioritisation of the terrestrial. The exhibition *Sea Inside* seeks to disrupt this tendency.

Artists have troubled the museum as a site of oceanic knowledge, for instance in Brian Jungen's (b.1970) articulated whale skeletons, which appear like natural history specimens but are in fact made from white plastic garden chairs. This sublimation of cheap commercial products into museum objects elicits questions about value and the formation of knowledge, both in the Western tradition and – with titles like *Shapeshifter* (2000) – in relation to Indigenous cosmologies. The longstanding photographic inquiry into natural history dioramas by Hiroshi Sugimoto (b.1948), which began in 1976, draws attention to the epistemology of the museum through a focus on the crafted artificiality of these stylised 'windows on nature' and their attempts to arrest wildlife in a moment in time.[42] *Devonian Period* (1992/2004) presents a Palaeozoic seabed scape in which early marine flora and fauna can be seen depicted millions of years before the advent of humans, as if captured within the glass of an aquarium, or even alive in their original habitat (fig. 11.6). Such a vision reminds viewers of the primordial character of oceans, which called forth the origins of life as we know it and from which we humans all evolved, reiterating once again this intrinsic interconnection that humans have with the sea, but also its inevitable mediation.

It is this sense of continuity of humans with oceans and ocean life, as well as the structures of containment and connection that the exhibition *Sea Inside* explores through multimedia artworks across moving image, sculpture, photography and works on paper, but also with lineages grounded in the decorative arts, craft, fashion and jewellery. The exhibition seeks to highlight our more intimate and imaginative relationships with the sea and marine life that might foster an ethical sensibility in the face of the destruction of ocean species and habitats, to advocate for new ways of thinking about our relations to the oceans on which our and others' lives depend.

'THE GHOST WHALES'

HARUN MORRISON

Notes for 'The Ghost Whales'

The *grindadráp* (or grind) is a tradition of killing pilot whales and sometimes other dolphins dating back to 800 CE. Until relatively recently, pilot whale meat was an important source of food, but there is no longer a need for whale meat to meet nutritional needs. *Grindadráp* (from the Faroese terms *grindhvalur*, meaning pilot whale, and *dráp*, meaning killing), is a type of drive hunting that involves herding various species of whales and dolphins, but primarily pilot whales, into shallow bays to be beached, killed and butchered. Each year, an average of around 700 long-finned pilot whales and several hundred Atlantic white-sided dolphins are caught over the course of the hunt season during the summer. The *Ecologist* magazine has reported on the Royal Danish Navy protecting this ritual by sending warships to disrupt potential marine-based protests and interventions like Sea Shepherd:

> *There was no Danish intervention in the years prior to 2014. They did not send their warships in the past. They are doing so now. Even though killing whales is illegal under European Union law, the government of Denmark has thrown their weight behind the killers.*[1]

The lyrics for 'The Ghost Whales' overwrite those of 'The Last Shanty' by Tom Lewis.

'THE GHOST WHALES'

Melody: 'The Last Shanty' by Tom Lewis, 1980
Lyrics commissioned by Hosting Lands, 2024

[1] They corralled the pilot whales into the shallow harbour
Men came running from the shore with spears and hollow laughter
It wasn't long before the sea was turned from blue to red
The tide came out, a single hunt, a thousand whales lay dead.

Refr. Whales swimming backwards, lances fly to hands,
Dreaming of reversal, of sun-kissed dorsal fins
The final drops of blood revive their bodies' veins.
And those who came for carrion can carry on in vain.

[2] Whales' mournful courting songs are now replaced by cries
The navy oversaw this carnage under grieving skies,
I understand this ritual has run for centuries
But when do we decide to give up certainties?

Refr. Whales swimming backwards, lances fly to hands ...

[3] We hear human reactions, voices back and forth
Debates from all positions, west, east, south and north
Still the sea stays silent, does it recognise these horrors?
The only break of quiet lies in its rippling mirrors.

Refr. Whales swimming backwards, lances fly to hands ...

[4] Old sailors recount ghost ships drifting through the night
But have you faced a ghost whale pod cruising in moon light?
The ghost whales were not silent we recognised their song
They encircled us each sun-set reminding what was done.

Refr. Whales swimming backwards, lances fly to hands ...

[5] As the night descended life aboard became more strange
Whalers would clasp their ears as though escaping a gun range
Men would twist and turn their backs as though their spines were ripped
And when they opened mouths to scream their tongues were in a grip.

Refr. Whales swimming backwards, lances fly to hands ...

[6] When the men came ashore no one took their word
And when they questioned future hunts, their protests were unheard
This crew were treated see-through by their neighbours in their streets
The sailors soon began to doubt whether they exist.

ACKNOWLEDGEMENTS

Can the Seas Survive Us? marks a pivotal milestone in the Sainsbury Centre's commitment to envisioning a sustainable future through cultural and ecological lenses – a commitment strengthened since the Centre's relaunch in 2023. This project has taken us on a journey, both physical and metaphorical: from a two-day walk from the Sainsbury Centre to Great Yarmouth, to a journey across the North Sea on a century-old fishing vessel, exploring oceanic landscapes, the industrialisation of the North Sea, and East Anglia's historical ties with Dutch coastal communities. This innovative curatorial research, exhibition programme and book would not have been possible without the advice, collegiality and generosity of many people.

We thank the artists, writers, activists and researchers whose diverse perspectives enrich this volume with critical essays that deepen our understanding of art's role in fostering and accelerating discussions around cultural-environmental issues.

This project has involved many partners, including at Norwich University of the Arts, Louis Nixon, Candice Alison and Teresa Stopanni. Artists Ed Compson and Arieh Frosh and Kaavous Clayton and Jules Devonshire of Original Projects, Great Yarmouth sailed with us to the Netherlands and met artists, including Jan Eric Visser, Nabuurs&VanDoorn, de Onkruidenier, Annabel Howland and Boris Maas, who made our research trip unforgettable. The Excelsior Trust voyaged with us across the North Sea. Partners in the Netherlands included the Rotterdam Architecture Biennale team, Yonca Ozbilge, Katherine Koekoek, Saskia van Stein and Hani Salih; and the inspiring Nieuwe Instituut team including Aric Chen, Francien van Westrenen, Marie-Anne Soulemaic and Tijn van de Wijdeven, with special recognition to Suzanne Mulder, Isabelle van der Riet and Elza van den Berg.

The editors extend their thanks to everyone involved in realising the exhibitions *A World of Water*, *Darwin in Paradise Camp: Yuki Kihara* and *Sea Inside*, alongside related interventions and programmes, including George Nuku's sensitive interaction with the cases that house the Māori objects in the Sainsbury Centre's collection display. Yuki Kihara worked tirelessly to execute her exceptional ideas to adapt *Paradise Camp* for the Sainsbury Centre. Pandora Syperek and Sarah Wade brought their expert insight and knowledge to the curation of *Sea Inside*.

A World of Water was curated by John Kenneth Paranada and *Darwin in Paradise Camp: Yuki Kihara* by Tania Moore. We thank the Whitworth Art Gallery for partnering with us on *Darwin in Paradise Camp: Yuki Kihara*, especially Darren Pih and Poppy Bowers. The Natural History Museum offered great support in the development of Yuki Kihara's research around Charles Darwin, particularly James Maclaine.

This publication and exhibition programme would not have been possible without the generous support of the artists and their studios, galleries and museum partners. Lenders include the Smithsonian National Museum of American History; the Barber Institute of Fine Arts; the Courtauld Institute of Art; National Trust Collections; Government Art Collection; Tyler Eash; Chioma Ebinama; Time and Tide Museum of Great Yarmouth Life; Evan Ifekoya; The Fitzwilliam Museum, University of Cambridge; Gabriella Hirst; Royal Academy of Arts; Norfolk Archive Office; Nieuwe Instituut; Museum of Archaeology and Anthropology, University of Cambridge; Migros Museum für Gegenwartskunst; Natural History Museum; Keith and Rhonda Lloyd; Nome Gallery; Helen Sunderland-Cohen of The Sunderland Collection; Debbie Carslaw; Nessie Stonebridge; I8 Gallery; Studio Julian Charrière; Marcus Coates, Alexander Grover of Modern Art Gallery; Richard Saltoun Gallery; Maggi Hambling's Studio; Julian Perry; Laure Provost; Skye Turner; El Morgan; Claire Cansick; Jayne Ivimey; Cian Dayrit; Amanda Wilkinson Gallery; Yuki Kihara; Alexander Vourecas Petalas; and all the artists in the exhibitions. We are also grateful to John Packman and Andrew Farrell from the Broads Authority for their advice and insights, as well as the numerous researchers with whom we've collaborated. We also thank Gui Taccetti, Yuki Kihara's production manager for his tireless work in bringing her exhibition to fruition. Some of Yuki Kihara's *Paradise Camp* series will remain in the Sainsbury Centre's collection thanks to generous private funding and the support of Keith and Rhonda Lloyd, Paul and Gillian Kendrick, and Te Kano Estate.

The research and engagement for Yuki Kihara's exhibition was made possible by the AHRC IAA Strategic Themes Funding, administered by UEA and awarded to Karen Jacobs. The British Council's Connections Through Culture grant generously allowed for the production of Kihara's new work Darwin Drag as well as a public talk at the Sainsbury Centre. Creative New Zealand has supported Kihara's practice, allowing her to develop the project for the Sainsbury Centre. The Dutch Embassy have generously supported *A World of Water*, and the exhibition's research in the Netherlands was generously supported by a Jonathan Ruffer grant from the Art Fund.

This season was made possible by the pioneering role of John Kenneth Paranada, appointed Curator of Art and Climate Change in 2022 – a post generously funded by the John Ellerman Foundation, whose dedication to sustainability in the arts enabled this ambitious endeavour. Our appreciation also extends to the Embassy of the Kingdom of the Netherlands in London for their valued support, including Ambassador Paul Hujits, Cultural Attaché Astrid de Vries, Cultural Policy Officer Koen Guiking and Andrew Wood, Hon Dutch Consul for the East of England. The British Council's Netherlands Country Director, Jennifer Cosgrave, has been instrumental in forging future partnerships with Dutch cultural institutions, including the Kröller-Müller Museum, Maritiem Museum and Nieuwe Instituut, strengthening historical ties and promoting collaborative cultural initiatives between the UK and the Netherlands.

We are grateful to the entire Sainsbury Centre team who worked tirelessly to deliver the season and this book. Furthermore, the University of East Anglia's Tyndall Centre for Climate Change Research has been instrumental to this project. Executive Director Asher Minns, alongside Robert Nicholls and Andrew Watkinson, collaborated closely with John Kenneth Paranada, whose integration of scientific insights was invaluable in creating an exhibition that resonates with contemporary climate issues. We thank the Sainsbury Research Unit, particularly Karen Jacobs and Pat Hewitt.

We extend our sincere appreciation to the UEA Climate Narratives researchers, particularly Jean McNeil, Jos Smith, Steven Waters, Rebecca Tillet, Joanne Clarke, Christine Cornea, Tom Greaves and the numerous distinguished scholars whose contributions have shaped the curatorial vision for this season. Their pioneering work in the emerging field of Climate Narratives, in collaboration with our Curator of Art and Climate Change, has positioned UEA as a leader in interdisciplinary research, integrating creative and scientific practices to advance a sustainable future.

We also wish to express our gratitude to the Mile Cross community for their openness and deep engagement with climate action – a testament to the power of collective resilience. Additional thanks are extended to James Wreford and Ned Davies of Helgate Pottery, Lucy Allen of Norfolk and Waveney Mind, as well as Judy Omaste of the Norwich City Council Community Services for their invaluable support and involvement. This collaboration has been made possible through the dedicated efforts of Alexander Bratt of Creative UEA and Karina Aveyard of the Green Film Festival, whose support has been instrumental in bringing this project to fruition.

Our warmest appreciation goes to Lizzy Silverton and James Alexander for their meticulous attention to editing and design. We are grateful to the photographers for visually encapsulating the exhibition's themes and to Kulturalis for bringing this publication to print. Hudson Architects, led by Jack Spencer Ashworth, crafted an innovative design for *A World of Water* and *Sea Inside*, repurposing materials from the Centre's stores to create a sustainable, low-impact aesthetic evocative of the marine depths. Anita Gigi transformed Yuki Kihara's *Paradise Camp* for the Sainsbury Centre. George Sexton Associates carried out expert lighting design and focusing across the season.

Finally, to every reader and visitor engaging with this work: we hope that this exhibition and publication inspire meaningful reflection on our shared responsibility to protect the seas and the delicate ecosystems that sustain us all. Thank you for joining us on this journey.

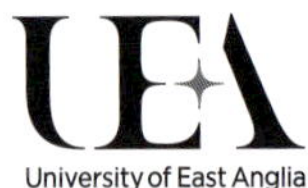

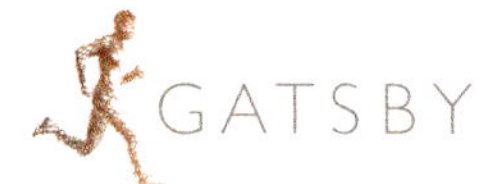

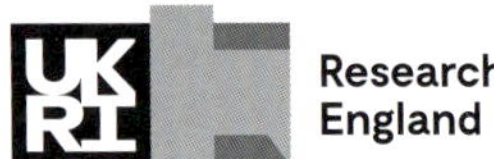

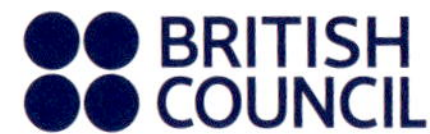

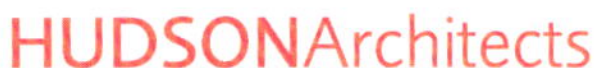

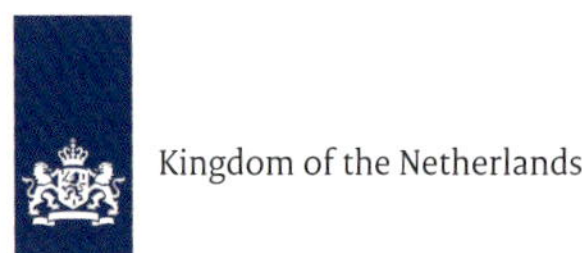

AUTHOR BIOGRAPHIES

Antonia Blocker

Antonia Blocker is a writer and curator interested in artists who create in disruptive, imaginative and expansive ways, and who are not well supported by traditional exhibition structures. From 2024 to 2025, she was Head of Exhibitions at Modern Art Oxford. Prior to this, she was Senior Curator at the Zabludowicz Collection, where she led on bringing adventurous and experimental practices to the UK, while supporting early career artists and art professionals nationally. During this time, she commissioned projects by Lu Yang, Tianzhuo Chen, Trulee Hall and Fani Parali, among others. She has held curatorial positions at the Whitechapel Gallery, ICA and Serpentine Galleries and has curated projects at Gasworks and The Showroom. Internationally, she has curated projects at LOOP Barcelona and taken part in discussions at Performance Affair in Brussels. Most recently, her writing has been published in *Art Monthly* and on the Performance Art Museum website. She has been a trustee for a decade, formerly at Lighthouse in Brighton and currently at Stanley Arts in Croydon.

Soren Brothers

Soren Brothers is the Allan and Helaine Shiff Curator of Climate Change at the Royal Ontario Museum, Toronto. He is also an Assistant Professor at the Department of Ecology and Evolutionary Biology at the University of Toronto. Soren's research examines the effects of climate change on lakes, and how fluctuations in aquatic systems can influence their greenhouse gas emissions into the atmosphere. More broadly, he is interested in understanding how feedback loops and the transdisciplinary study of lakes can help us better understand and predict global tipping points that may accelerate anthropogenic climate change.

Born in Mississauga and raised in Toronto, Soren has worked on lakes in a diverse array of environments around the world, including the Nunavut tundra, Quebec's boreal forests and the Great Lakes. He is also passionate about science communication and community outreach. Before beginning at the Royal Ontario Museum in 2021 he was an Assistant Professor of Limnology at Utah State University and a CREATE programme manager and postdoctoral fellow at the University of Guelph, focusing on multiple stressors and cumulative effects in the Great Lakes.

Jago Cooper

Jago Cooper is Executive Director of the Sainsbury Centre and Professor of Art and Archaeology at the University of East Anglia. After joining his local museum as a six-year-old volunteer in the 1980s, he started his first job at the Museum of London in the 1990s before spending a decade living and working in Sri Lanka, Australia and Cuba in museum and cultural heritage roles. After returning to the UK to teach at the University of Leicester and UCL, Jago was appointed as the Head of the Americas Section by the British Museum where he enjoyed creating a series of major exhibitions, new digital experiences and international museum collaborations including the establishment of the Santo Domingo Centre of Excellence for Latin American research.

Jago's numerous books, articles and television documentaries reflect his interest in tackling the big questions of our future from climate change to technological innovation. His recent publications include *Living Art Sharing Stories* (2023), *Mapping a New Museum* (2022), *Peru: A Journey in Time* (2021) and *Arctic: Culture and Climate* (2020). He took up his role at the Sainsbury Centre in 2021, which, under his leadership, has undergone a major relaunch, becoming the first museum in the world to recognise the living lifeforce of art, and is now 'probably the UK's most radical art museum' (*Guardian*).

Patrick Flores

Patrick Flores is Professor of Art Studies at the University of the Philippines and concurrently Chief Curator of the National Gallery Singapore. He is also the Director of the Philippine Contemporary Art Network. He was a Visiting Fellow at the National Gallery of Art in Washington D.C. in 1999. Among his publications are *Painting History: Revisions in Philippine Colonial Art* (1999), *Past Peripheral: Curation in Southeast Asia* (2008), *Art After War: 1948–1969* (2015) and *Raymundo Albano: Texts* (2017). He was a Guest Scholar of the Getty Research Institute in Los Angeles in 2014, the Artistic Director of the Singapore Biennale 2019 and Curator of the Taiwan Pavilion at the Venice Biennale in 2022.

Karen Jacobs

Karen Jacobs is Associate Professor at the Sainsbury Research Unit, University of East Anglia. Her research focuses broadly on the arts from Oceania and specifically on museum anthropology and related debates, aspects of climate change and ocean culture, missionary heritage, body adornment, youth culture and contemporary Pacific art. Her research is in collaboration with museums, Indigenous communities and contemporary artists. Most of her research has been conducted in the framework of funded international research projects and has resulted in a variety of exhibitions and publications. Book projects include *Collecting Kamoro* (2012) and *This is Not a Grass Skirt* (2019). Exhibition projects include *Pacific Encounters* (Musée du quai Branly – Jacques Chirac, 2008) and *Fiji: Art and Life in the Pacific* (Sainsbury Centre, 2016–17; Los Angeles County Museum of Art, 2019–21).

Yuki Kihara

Yuki Kihara is an award-winning artist and curator/producer whose art practice is characterised by a research-driven, interdisciplinary approach. Her work challenges the dominant and singular historical narrative through a wide range of media, including performance, sculpture, film, photography and curatorial practice. Yuki has exhibited in over 70 institutions worldwide including representing Aotearoa New Zealand at the 59th Venice Biennale in 2022. Her works are held in over 30 permanent collections internationally, including the Museum of Modern Art, New York, and she has curated a number of exhibitions including at Carriageworks, Sydney and the Bernice Puhiwai Bishop Museum, Honolulu. Yuki has received over 50 grants, awards and sponsorships, these include the National Museums of World Cultures, the Netherlands; the Arts Council of New Zealand Toi Aotearoa; and the Government of Sāmoa.

Yuki is an affiliate of Ecological Art Practices – a research cluster led by THE NEW INSTITUTE Centre for Environmental Humanities (NICHE) at the Ca 'Foscari University of Venice.

Yuki lives and works in Sāmoa.

Tania Moore

Tania Moore is Head of Exhibitions at the Sainsbury Centre. She is co-editor of *Can the Seas Survive Us?* and worked with Yuki Kihara to conceive *Paradise Camp* for the Sainsbury Centre's *Can the Seas Survive Us?* season. Previously, she curated *Lindsey Mendick: Hot Mess*, *In Event of Moon Disaster*, *Liquid Gender* and *Jeffrey Gibson: no simple word for time* (all Sainsbury Centre, 2024). Publications include *What Is Truth?* (2024), *Rhythm and Geometry: Constructivist Art in Britain Since 1951* (2021) and *Henry Moore: Friendships and Legacies* (2020). She has contributed chapters on post-war and contemporary sculpture in publications including *Magdalene Odundo: The Journey of Things* (2019), *Elisabeth Frink: Humans and Other Animals* (2018) and *Becoming Henry Moore* (2017). In 2019 she received the New Collecting Award from the Art Fund to acquire sculptors' drawings by contemporary women and non-binary artists for the Sainsbury Centre collection.

Harun Morrison

Harun Morrison is an artist and writer based in London and an associate artist with Greenpeace UK on the project *Bad Taste*. In the summer of 2024 he was part of a two-person show, *DONO*, at Somerset House Studios project space G31 alongside Appau Jnr Boakye-Yiadom. His latest group exhibitions include the *Sonic Acts Biennial 2024: The Spell of the Sensuous* (Amsterdam, 2024), *Chronic Hunger, Chronic Desire* (Timișoara, 2023), *BALATORIUM: Disturbed Waters* (Veszprém, 2023) and *Storm Warning: What does climate change mean for coastal communities?* (Focal Point Gallery; Newlyn Art Gallery, 2023–4). Recent solo exhibitions include *Dolphin Head Mountain* (Horniman Museum, 2022–3), *Mark the Spark* (Nieuwe Vide, 2022) and *Experiments with Everyday Objects* (Eastside Projects, 2021). Harun is currently co-developing community gardens in Merseyside for the Bootle Library and Mind Sheffield, a mental health support service, as part of the Arts Catalyst research project *Emergent Ecologies*. His forthcoming novel, *The Escape Artist*, will be published by Book Works in 2025.

Djoeke van Netten

Djoeke van Netten is Associate Professor of Early Modern History at the University of Amsterdam. Her research is at the crossroads of the history of knowledge, maritime history, and the history of printed books and maps, with a focus on the Dutch Republic in a global context. Djoeke's PhD from the University of Groningen addressed the role of printers and publishers in the presentation and dissemination of science (published as *Koopman in Kennis*, 2014). She led on the project *Hide and Leak* on secrecy and openness in the early Dutch overseas companies, funded by the Dutch Research Council (NWO). Over the last few years, she has edited dedicated issues on women at sea, the circular economy in early modern times and mapping uncertain knowledge. Her recent research focuses on mapping oceans, mapping ice and on animal history, especially polar bears. At the University of Amsterdam, Djoeke teaches world history, military history and the history of cartography. She is affiliated with the National Maritime Museum in Amsterdam and is a member of the editorial board of the *Dutch Journal for Maritime History*, the board of the Dutch–Flemish Society for Early Modern History and the Linschoten-Vereeniging.

John Kenneth Paranada

John Kenneth Paranada is co-editor of *Can the Seas Survive Us?* and curator of *A World of Water* as part of the related season. He developed *Towards the Weird Heart of Things*, a commission with Ivan Morison for the Sainsbury Centre Sculpture Park and Orleans House Gallery for the *Why Do We Take Drugs?* season in 2024. He serves as the Curator of Art and Climate Change at the Sainsbury

Centre, funded by the John Ellerman Foundation, and is a researcher at the Tyndall Centre for Climate Change Research, University of East Anglia. His interdisciplinary expertise spans museum studies, curatorial studies, art history, community engagement, climate communication and energy management. In 2023 he curated *Sediment Spirit* and co-edited the publication *Planet for Our Future*. Recent publications include: 'A Path Forward: Curating Art and Climate Change at the Sainsbury Centre', in *Museum International* (2024), 'How Do We Begin a Meaningful Conversation About Art's Place in the Climate Crisis', in *Design for Our Planet* (2023) and 'Collisions: Art & Climate Change', in *Adaptation: A Reconnected Earth* (2023).

Pandora Syperek

Pandora Syperek is co-curator of the exhibition *Sea Inside*, which runs as part of the *Can the Seas Survive Us?* season at the Sainsbury Centre. She is Leverhulme Early Career Fellow at Loughborough University London and Visiting Fellow at the V&A Research Institute. She researches the intersections of art, science, gender and the nonhuman within cultures of display. She is co-investigator on the Paul Mellon Centre-funded collaborative research project *Exhibiting Oceans in the UK Today*, and co-lead on Loughborough's Institute of Advanced Studies' 2023–4 annual theme, *Gestation: Bodies, Technologies, Ecologies, Justice*. She co-edited the anthology *Oceans* (*Documents of Contemporary Art*) (2023) and the *Journal of Curatorial Studies* special issue 'Curating the Sea' (2020), and has published book chapters and journal articles on gender and ecological aesthetics in natural history display as well as texts on contemporary art. Pandora was a postdoctoral fellow at the Paul Mellon Centre from 2016 to 2017 and she holds a PhD in the History of Art from University College London (UCL), funded by SSHRC. She has taught on modern and contemporary art, design and curatorial studies at Loughborough University, Sotheby's Institute of Art, UCL and the University of York, and previously worked on the curatorial team at the Banff Centre, Canada.

Courtney Traub

Courtney Traub is a Norwich-based non-fiction writer and editor who has authored essays on the history of Romanticism, literature, environment and technology in publications including *Arizona Quarterly*, *Studies in the Novel* and *Literature Compass*. She holds a Doctorate in English from the University of Oxford, and is currently working on a non-fiction book that explores the cultural and historical connections between nineteenth-century Romantic and Transcendentalist thought, countercultural technology and ecology movements of the twentieth century, and how we conceptualise or respond to contemporary ecological and technological crises. In addition to her scholarly work, Courtney is a copy-editor for an international museum studies journal and a travel writer with expertise on French and Parisian history and culture.

Sarah Wade

Sarah Wade is co-curator of *Sea Inside* as part of *Can the Seas Survive Us?* She is an art historian and Associate Professor in Museum Studies at the University of East Anglia. Her research focuses on human–animal relations and representations of wildlife in contemporary art, exhibitions and museum displays particularly with regards to ecological concerns. Sarah holds a PhD in the History of Art from University College London and has published widely on extinction and wildlife conservation issues in artistic and curatorial practice. She is co-founder of the *Curating the Sea* research project, which resulted in a special issue of the *Journal of Curatorial Studies* (2020), *Oceans: Documents of Contemporary Art* (2023) and numerous other events and publications including the *Artist/Oceans* series (2022) and *Connecting Oceans in Conversation* (2024) with contemporary artists and arts organisations. Over the years, Sarah has worked with various museums and heritage organisations in research, project management and curatorial capacities, and continues to collaborate with museum colleagues on research projects and exhibitions.

Andrew Watkinson

Andrew Watkinson is an Emeritus Professor of Environmental Sciences at the University of East Anglia and, until recently, chair of Forest Research's Expert Committee on Forest Science, a Non-Executive Director of the Centre for Environment, Fisheries and Aquaculture Science (Cefas) and Vice President of the British Trust for Ornithology. Previously he was Director of the Natural Environment Research Council's programme Living With Environmental Change and the Tyndall Centre for Climate Change Research, where he took a particular interest in flooding and the impact of climate change on the coastal zone. He contributed to the award of the Nobel Peace Prize for 2007 to the Intergovernmental Panel on Climate Change and was awarded the Marsh Prize for Ecology by the British Ecological Society. An ecologist by background, Andrew has worked increasingly on the urgent topic of environmental change, taking a broad interdisciplinary perspective and working with natural scientists, engineers and social scientists. His interest in art and the environment began with his involvement in the work of the GroundWork Gallery in King's Lynn and has since developed in research on the Norwich School Artist James Stark and collaborations with the environmental artist Jayne Ivimey.

PICTURE CREDITS

Every effort has been made to acknowledge correct copyright of images where applicable. Any errors or omissions are unintentional and should be notified to the Publisher, who will arrange for corrections to appear in any future editions.

Fig. 0.0 Commissioned by East Gallery, Norwich University of the Arts
Figs 0.1, 0.2, 0.6, 0.13, 5.4 © Julian Charrière; VG Bild-Kunst, Bonn, Germany
Fig. 0.3 © Koen Taselaar
Fig. 0.4 © Olafur Eliasson; photograph Vigfús Birgisson
Figs 0.5, 6.3 © Claire Cansick
Fig. 0.7 Courtesy of Arieh Frosh and Ed Compson; commissioned by East Gallery, Norwich University of the Arts
Fig. 0.8 © Arieh Frosh and Ed Compson
Figs 0.9, 0.11 © De Onkruidenier; photographs Marleen Annema
Fig. 0.10 © Nabuurs&VanDoorn
Figs 0.12, 3.3 The Fitzwilliam Museum, University of Cambridge
Fig. 1.1 Courtesy of Google Maps
Figs 1.2, 1.3, 1.4 Courtesy of The Sunderland Collection
Fig. 1.5 Courtesy of the British Library, London
Figs 2.1, 2.2, 2.3, 2.5 Courtesy of Soren Brothers
Fig. 2.4 Courtesy of ROM (Royal Ontario Museum), Toronto, Canada. Mike Tijoe © ROM
Fig. 3.1 Courtesy of Norfolk Record Office, Y/C 37/1
Fig. 3.2 Photograph Andrew Watkinson
Fig. 3.4 Courtesy of Time and Tide Museum of Great Yarmouth Life
Figs 3.5, 5.2 Courtesy of Tate, London
Fig. 3.6 © Jayne Ivimey
Fig. 3.7 © Julian Perry
Fig. 4.1 Courtesy of Harun Morrison
Fig. 5.1 Courtesy of Hamburger Kunsthalle, Hamburg, Germany
Fig. 5.3 © Bonnie Monteleone
Fig. 5.5 Courtesy of Flowers Gallery, New York
Figs 6.1, 6.2 © Maggi Hambling; photographs Doug Atfield
Fig. 6.4 Courtesy of the Government Art Collection
Fig. 8.1 © Pam Debenham. Courtesy of the National Gallery of Australia, Canberra
Fig. 8.2 © Jane Chang Mi
Fig. 8.3 © The artists. Courtesy of Auckland War Memorial Museum Tāmaki Paenga Hira
Fig. 8.4 Courtesy of Mairie de Cannes
Fig. 8.5 © DACS 2024. Courtesy of the Museum of Archaeology and Anthropology, Cambridge
Fig. 8.6 Courtesy of Sullivan+Strumpf, Zetland, Australia
Figs 8.7, 9.1, 9.2, 9.3, 9.4 © Yuki Kihara
Fig. 9.2 Exhibition still by Ralph Brown
Figs 10.1, 10.2 Courtesy of the Library of Congress, Washington, D.C.
Figs 10. 3, 10.4 Courtesy of NCCA-PAVB; photographs Jeanne Melissa Severo
Fig. 10.5 Courtesy of Bangko Sentral ng Pilipinas Collection
Fig. 10.6 Produced by LVN Pictures, Inc.; screenshot courtesy of ABS-CBN Film Restoration (Sagip Pelikula)
Fig. 11.1 © The Estate of Ana Mendieta Collection, LLC. Licensed by ARS, New York/ DACS, London. Courtesy Galerie Lelong & Co.
Fig. 11.2 © Evan Ifekoya. All rights reserved, DACS/Artimage 2024
Fig. 11.3 © Laure Prouvost. Courtesy Lisson Gallery
Fig. 11.4 © Chioma Ebinama; image courtesy of Maureen Paley, London
Fig. 11.5 © Shuvinai Ashoona; image courtesy Feheley Fine Arts and Dorset Fine Arts
Fig. 11.6 © Hiroshi Sugimoto. Courtesy Lisson Gallery

NOTES

Introduction

1 The Kyoto Protocol, adopted in 1997 and entered into force in 2005, is an international treaty that commits industrialised countries to reduce greenhouse gas emissions based on agreed targets. It represents a crucial step in global efforts to combat climate change. United Nations Framework Convention on Climate Change (UNFCCC), *Kyoto Protocol to the United Nations Framework Convention on Climate Change*, 1997, pp. 1–20.

2 The Paris Agreement, adopted in 2015, is a global treaty under the UNFCCC aiming to limit global warming to well below 2°C, with efforts to stay below 1.5°C. It encourages countries to set national targets to reduce greenhouse gas emissions and strengthen resilience to climate impacts. United Nations Framework Convention on Climate Change (UNFCCC), *The Paris Agreement*, 2015, pp. 1–32.

3 The United Nations Sustainable Development Goals (SDGs) for 2030 are a global framework of 17 goals adopted in 2015, aiming to end poverty, protect the planet, and ensure prosperity for all by 2030. They address issues like climate change, inequality and environmental sustainability. United Nations, *Transforming Our World: The 2030 Agenda for Sustainable Development*, 2015, pp. 1–35.

4 Sustainable Development Goal 14 (SDG 14) focuses on conserving and sustainably using the oceans, seas and marine resources. It aims to reduce marine pollution, protect ecosystems, and ensure sustainable fishing practices by 2030, addressing critical issues like ocean acidification and overfishing. United Nations, *The Sustainable Development Goals Report 2020*, 2020, pp. 54–9.

5 The IPCC's COP26, held in Glasgow in 2021, also highlighted the role of heritage and art in climate action. Cultural institutions and artists were integral in raising awareness of climate change impacts, emphasising the need to protect both natural and cultural heritage from environmental degradation. United Nations, *COP26: The Glasgow Climate Pact*, 2021, pp. 1–97. At COP27, held in Sharm El-Sheikh in 2022, the role of museums and heritage sites in climate action was emphasised. These institutions were recognised as key players in educating the public on climate change, preserving cultural heritage, and fostering global awareness through exhibitions and community engagement. United Nations, *COP27: Sharm El-Sheikh Implementation Plan*, 2022, pp. 1–60.

6 Sylvia Earle, *The World Is Blue: How Our Fate and the Ocean's Are One* (National Geographic, 2009), p. 6.

7 Michael Pye, *The Edge of the World: How the North Sea Made Us Who We Are* (Penguin Books, 2015), pp. 1–394.

8 Hal Harvey, Robbie Orvis and Jeffrey Rissman, *Designing Climate Solutions: A Policy Guide for Low-Carbon Energy* (Island Press, 2018), pp. 1–272.

Fluid Lines: Charting Change and Changing Charts

1 I would like to thank Ken Paranada (Sainsbury Centre) and Helen Sunderland (The Sunderland Collection). Moreover, I am very grateful to Nick Millea and Jaap van Netten for their valuable comments. Thanks as well to beach pavilions Rapa Nui and Noosa for coffee and sea views.

2 Mark Monmonier, *Coast Lines: How Mapmakers Frame the World and Chart Environmental Change* (Chicago University Press, 2008).

3 Visualising the coastline paradox will clarify even better. YouTube offers various insightful animations.

4 J. B. Harley and David Woodward, 'Preface', in *The History of Cartography*, vol. 1 (Chicago University Press, 1987), p. xvi.

5 Lewis Carroll, *Sylvie and Bruno Concluded* (Macmillan and Co., 1893) chapter XI; Umberto Eco, *How to Travel with a Salmon and Other Essays* (Harcourt, 1994).

6 J. B. Harley, 'Silences and Secrecy: The Hidden Agenda of Cartography in Early Modern Europe', *Imago Mundi*, (1988) vol. 40, no. 1, pp. 57–76.

7 Mark Monmonier, *Rhumb Lines and Map Wars: A Social History of the Mercator Projection* (Chicago University Press, 1994).
8 Drag countries around to compare their size on https://www.thetruesize.com/.
9 Catherine Delano-Smith, 'Signs on Printed Topographical Maps, *c.*1470–*c.*1640', in David Woodward (ed.), *The History of Cartography,* vol. 3.1 (Chicago University Press, 2007), p. 542.
10 Rodney Shirley, *The Mapping of the World: Early Printed World Maps, 1472–1700* (Early World Press, 2001); Cornelis Koeman, *Atlantes Neerlandici: Bibliography of Terrestrial, Maritime and Celestial Atlases and Pilot Books, published in the Netherlands up to 1880*, vol. IV (Theatrum Orbis Terrarum, 1970).
11 Athanasius Kircher, *Mundus Subterraneus* (Janssonius, 1665).
12 Delano-Smith, 'Signs on Printed Topographical Maps', p. 542.
13 Djoeke van Netten, 'The New World Map and the Old: The Moving Narrative of Joan Blaeu's *Nova Totius Terrarum Orbis Tabula* (1648)', in Zef Segal and Bram Vannieuwenhuyze (eds), *Motion in Maps, Maps in Motion: Mapping Stories and Movement through Time* (Amsterdam University Press, 2020), pp. 33–56.
14 Jordana Dym, *Mapping Travel: The Origins and Conventions of Western Journey Maps* (Brill, 2021).
15 Delano-Smith, 'Signs on Printed Topographical Maps', p. 543.
16 There is not much literature on coastal profiles, except the unpublished MA thesis by Wouter de Vries (Free University, Amsterdam, 2018).
17 For (mostly) Portuguese examples, see the Rutter project, making the Earth global, https://rutter-project.org [accessed 10 September 2024].
18 On this genre see Djoeke van Netten, *Koopman in Kennis: De Uitgever Willem Jansz Blaeu in de Geleerde Wereld 1571–1638* (Walburg Pers, 2014), chapter 2.
19 Dagomar Degroot, *The Frigid Golden Age: Climate Change, the Little Ice Age, and the Dutch Republic, 1560–1720* (Cambridge University Press, 2018).
20 María Portuondo, *Secret Science: Spanish Cosmography and the New World* (Chicago University Press, 2009).
21 Kees Zandvliet, *Mapping for Money: Maps, Plans and Topographic Paintings and their Role in Dutch Overseas Expansion during the Sixteenth and Seventeenth Centuries* (Batavian Lion International, 1998).
22 Lewis Carroll, *The Hunting of the Snark: An Agony in Eight Fits* (Macmillan and Co., 1876).
23 Lawrence Buell, *The Environmental Imagination: Thoreau, Nature Writing and the Formation of American Culture* (Harvard University Press, 1996); Avi Brisman, *Climate Change as a Crisis of Imagination* (Bristol University Press, 2025).

Sounding the Unknown: Waters and Global Change

1 Rita Adrian et al., 'Lakes as Sentinels of Climate Change', *Liminology and Oceanography* (2009), vol. 54, pp. 2283–97.
2 Soren Brothers et al., 'Declining Summertime pCO_2 in Tundra Lakes in a Granitic Landscape', *Global Biogeochemical Cycles* (2021), vol. 35, pp. 1–14; Kerri Finlay et al., 'Decrease in CO_2 Efflux from Northern Hardwater Lakes with Increasing Atmospheric Warming', *Nature* (2015), vol. 519, pp. 215–18.
3 Kevin C. Rose et al., 'Aquatic Deoxygenation as a Planetary Boundary and Key Regulator of Earth System Stability', *Nature Ecology & Evolution* (2024), vol. 8, doi:10.1038/s41559-024-02448-y.
4 M. Scheffer et al., 'Alternative Equilibria in Shallow Lakes', *Trends in Ecology & Evolution* (1993), vol. 8, pp. 275–9.
5 Malcolm T. McCulloch et al., '300 Years of Sclerosponge Thermometry Shows Global Warming has Exceeded 1.5 °C', *Nature Climate Change* (2024), vol. 14, pp. 171–7.
6 Devi Veytia et al., 'Overwinter Sea-ice Characteristics Important for Antarctic Krill Recruitment in the Southwest Atlantic', *Ecological Indicators* (2021), vol. 129, doi.org/10.1016/j.ecolind.2021.107934.
7 Terry P. Hughes et al., 'Spatial and Temporal Patterns of Mass Bleaching of Corals in the Anthropocene', *Science* (2018), vol. 359, pp. 80–83.
8 Ibid.; Michael J. Behrenfeld et al., 'Revaluating Ocean Warming Impacts on Global Phytoplankton', *Nature Climate Change* (2016), vol. 6, pp. 323–30.

9 Anny Cazenave and Frédérique Remy, 'Sea Level and Climate: Measurements and Causes of Changes', *Wiley Interdisciplinary Review Climate Change* (2011), vol. 2, pp. 647–62.
10 Timothy M. Lenton et al., 'Climate Tipping Points – Too Risky to Bet Against', *Nature* (2019), vol. 575, pp. 592–5.
11 Kristin Deasy, 'What We Know About the New High Seas Treaty', *npj Ocean Sustainability* (2023), vol. 2, pp. 1–3; Joachim Claudet, Cassandra M. Brooks and Robert Blasiak, 'Making Protected Areas in the High Seas Count', *Science* (2023), vol. 380, pp. 353–4.
12 Luís Gabriel A. Barboza et al., 'Macroplastics Pollution in the Marine Environment', *World Seas: An Environmental Evaluation Volume III: Ecological Issues and Environmental Impacts* (Elsevier Ltd., 2019). doi:10.1016/B978-0-12-805052-1.00019-X; Deepak Gola et al., 'The Impact of Microplastics on Marine Eenvironment: A Review', *Environmental Nanotechnology, Monitoring & Management* (2021), vol. 16, doi.org/10.1016/j.enmm.2021.100552.
13 Peter L. Macreadie et al., 'Blue Carbon as a Natural Climate Solution', *Nature Reviews Earth & Environment* (2021), vol. 2, pp. 826–39; Oswald J. Schmitz et al., 'Trophic Rewilding Can Expand Natural Climate Solutions', *Nature Climate Change* (2023), vol. 13, pp. 324–33.

Redrawing the Line

1 The origin of the name is obscure but probably derives from Dutch fishing vessels, Doggers, that used to fish the Dogger Bank. *Encyclopedia Britannica*, 13 February 2024, https://www.britannica.com/place/Dogger-Bank [accessed 18 August 2024].
2 Luc Amkreutz and Sasja van der Vaart-Verschoof (eds), *Doggerland: Lost World Under the North Sea* (Sidestone Press, 2022).
3 Brian Moss, *The Broads* (HarperCollins Publishers, 2001).
4 John Warden Robberds, *Geological and Historical Observations on the Eastern Vallies of Norfolk* (Bacon and Kinnebrook, 1826).
5 James Stark and John Warden Robberds, *Scenery of the Rivers of Norfolk* (Moon, Boys and Graves, 1834).
6 Andrew Moore, *The Norwich School of Artists* (HMSO, 1985).
7 Gareth H. H. Davies, *Great Yarmouth* (Poppyland Publishing, 2018).
8 Frank Meeres, *A History of Norwich* (Phillimore, 2016), p. 65.
9 Andrew Hemingway, *Landscape Imagery and Urban Culture in Early Nineteenth Century Britain* (Cambridge University Press, 1992), p. 209.
10 Richard J. Dawson et al., 'Integrated Analysis of Risks of Coastal Flooding and Cliff Erosion Under Scenarios of Long-term Change', *Climatic Change* (2009), no. 95, pp. 249–88.

'The Leaving of Kattegat Sea (or The Mermen's Lament)'

1 Miranda Bryant, 'Danish Village Under Threat From Landslide of Contaminated Soil', *Guardian*, 25 January 2024, https://www.theguardian.com/world/2024/jan/25/danish-village-threat-landslide-contaminated-soil-olst [accessed 10 September 2024].
2 See https://www.mariekoelbaek.com/ [accessed 25 November 2024].
3 Anders Galatius, 'Breeding Grey Seals are Still Rare in the Southern Baltic Sea, Danish Straits and Kattegat', *Wildlife Biology*, 12 October 2020, https://www.wildlifebiology.org/blog/breeding-grey-seals-are-still-rare-southern-baltic-sea-danish-straits-and-kattegat [accessed 10 September 2024].

'The Image of Eternity': Deep Time, Power and Vulnerability in Representations of Seas and Oceans

1 Lord Byron, 'Childe Harold's Pilgrimage', *The Complete Works of Lord Byron*, vol. 1. (W. Galignani and Co., 1842), p. 146.
2 'The Sea is History' was first published in *The Paris Review* in 1978. It later appeared in Walcott's collection *The Apple-Star Kingdom* in 1979.
3 Edmund Burke, *The Works of the Right Honourable Edmund Burke,* vol. I [1887] (Project Gutenberg, 2005), p. 131, https://www.gutenberg.org/files/15043/15043-h/15043-h.htm [accessed 11 July 2024].
4 Burke's was not the only notion of sublimity to gain popular traction in the late eighteenth century: Immanuel Kant's own conception is in many ways more subtle and complex. See

Immanuel Kant, *Critique of Judgement* [1790] (Oxford University Press, 2007) and Thomas Weiskel, *The Romantic Sublime: Studies in the Structure and Psychology of Transcendence* [1976] (Johns Hopkins University Press, 2019).

5 John P. Rafferty, 'Anthropocene Epoch', *Britannica*, https://www.britannica.com/science/Anthropocene-Epoch [accessed 8 August 2024].

6 See Courtney Traub, 'Ecocatastrophic Nightmares: Romantic Sublime Legacies in Contemporary American Experimental Fiction', *Arizona Quarterly* (Summer 2016), vol. 72, no. 2, pp. 29–60.

7 Timothy Morton, *Being Ecological* (Pelican Books, 2018), p. 22.

8 Samuel Taylor Coleridge, 'The Rime of the Ancient Mariner', in Samuel Taylor Coleridge, *Selected Poems*, Richard Holmes (ed.), (Penguin Books, 1996), p. 85.

9 Doré created the woodcuts for an 1877 German edition of Coleridge's lyrical poem, published under the translated title *Der alte Matrose.*

10 Coleridge, 'The Rime of the Ancient Mariner', p. 83.

11 'Toil and Terror at Sea', Tate, https://www.tate.org.uk/visit/tate-britain/display/jmw-turner/the-sea-toil-and-terror [accessed 1 August 2021].

12 From the gallery label for Great British Watercolors from the Paul Mellon Collection at the Yale Center for British Art, https://collections.britishart.yale.edu/catalog/tms:5458 [accessed 6 August 2024].

13 Byron, 'Childe Harold's Pilgrimage', p. 146.

14 Derek Walcott, 'The Sea is History', in *Collected Poems: 1948–1984* (Farrar, Straus and Giroux, 1986), p. 364.

15 John Perkins Marsh, *Man and Nature; or, Physical Geography as Modified by Human Action* (Sampson Low, Son and Marston, 1864), p. 534.

16 As Morton, op. cit., p. 64 argues, 'being able to understand durations is particularly important for us right now, because global warming's effects may last up to 100,000 years. What does that actually mean?'

17 'Introduction', Environment and Society Portal, (np), https://www.environmentandsociety.org/exhibitions/oceans-three-paradoxes/introduction [accessed 26 June 2024].

18 Nicole Block, 'Pretty Pollutants at Plastic Ocean Art Exhibition', *New University* (University of California, Irvine), newuniversity.org/2017/01/17/65303/ [accessed 4 August 2024].

19 Jane Bennett, *Vibrant Matter: A Political Ecology of Things* (Duke University Press, 2010).

20 Stefan Helmreich, 'Hokusai's Great Wave Enters the Anthropocene', *Environmental Humanities* (2015), vol. 7, no. 1, pp. 203–17, https://doi.org/10.1215/22011919-3616407 [accessed 1 August 2024].

21 Julien Charrière, *The Blue Fossil Entropic Stories*, https://julian-charriere.net/projects/the-blue-fossil-entropic-stories [accessed 9 August 2024].

22 Ibid.

23 '*Anthropocene* is a multidisciplinary body of work by Edward Burtynsky, Jennifer Baichwal and Nicholas de Pencier, which includes a photobook, a major travelling museum exhibition, a feature-length documentary film, and an interactive educational website.' https://www.edwardburtynsky.com/projects/photographs/anthropocene [accessed 12 August 2024].

Tipping Points: Pausing the Unstoppable in the Work of Maggi Hambling, Claire Cansick and Margaret Mellis

1 'Understanding climate tipping points', The European Space Agency, 6 December 2023, https://www.esa.int/Applications/Observing_the_Earth/Space_for_our_climate/Understanding_climate_tipping_points [accessed 25 July 2024].

2 Maggi Hambling, *Sea Sculpture: Paintings and Etchings* (Marlborough Fine Art, 2010).

3 Quoted from the press release for *Maggi Hambling: Real Time* at Marlborough New York, 10 March–30 April, 2022, https://www.marlboroughnewyork.com/attachment/en/5f46a387c8aa2caa468b4568/TextOneColumnWithFile/61b3d266e89a6807726a6d67 [accessed 25 July 2024].

4 Harriet Baker, 'Margaret Mellis: 1914–2009', *Margaret Mellis: Modernist Constructs* (The Redfern Gallery, 2021), p. 3.

5 Ibid.

6 Charles Taylor, *Cosmic Connections: Poetry in the Age of Disenchantment* (Harvard University Press, 2024), p. 130.
7 Astrida Neimanis, 'Hydro-Feminism: Or, On Becoming a Body of Water', in Henriette Gunkel, Chrysanthi Nigianni and Fanny Söderbäck (eds), *Undutiful Daughters: New Directions in Feminist Thought and Practice* (Palgrave Macmillan, 2012).
8 Ibid., pp. 92–3.
9 Damien Gayle, 'Dictionary Definitions: The Quest to Put the Human Back in Nature', *The Guardian*, 27 July 2024, p. 39.
10 Ibid.
11 Taylor, *Cosmic Connections: Poetry in the Age of Disenchantment*, p. 264.

A Sea of Resilience: Oceanic (In)visibility

1 Epeli Hau'ofa, 'Our Sea of Islands', *The Contemporary Pacific* (1994), vol. 6, no. 1, pp.148–61.
2 Epeli Hau'ofa, *We Are The Ocean: Selected Works* (University of Hawai'i Press, 2008), p. 31.
3 Ibid.
4 Albert Wendt, 'Towards a New Oceania', in Guy Amirthanayagam (ed.), *Writers in East-West Encounter: New Cultural Bearings* (Palgrave Macmillan, 1982), doi.org/10.1007/978-1-349-04943-1_12; Katerina Teaiwa, 'Reframing Oceania: Lessons from Pacific Studies', in Hilary E. Kahn (ed.), *Framing the Global: Entry Points for Research* (Indiana University Press, 2014); Teresia Teaiwa in Hau'ofa, *We Are The Ocean*, p. 41; Alice Te Punga Somerville, 'Where Oceans Come From', *Comparative Literature* (2017), vol. 69, no. 1, pp. 25–31.
5 Karl Chitham, Kolokesa Uafā Māhina-Tuai and Damian Skinner (eds), *Crafting Aotearoa: A Cultural History of Making in New Zealand and the Wider Moana Oceania* (Te Papa Press, 2019), p. 16.
6 Bronwen Douglas, 'Terra Australis to Oceania: Racial Geography in the "Fifth Part of the World"', *Journal of Pacific History* (2010), vol. 45, no. 2, pp. 179–210, doi.org/10.1080/00223344.2010.501696. Jules Sébastien César Dumont d'Urville, 'Sur les îles du Grand Océan', *Bulletin de la Société de Géographie* (1832), vol. 17, pp. 1–21. Dumont d'Urville also included Malaysia in his division, but throughout time Oceania has been considered to include Polynesia, Melanesia, Micronesia and Australia. These terms in themselves have been subjected to critical discussion (see, amongst others, Serge Tcherkézoff, 'A Long and Unfortunate Voyage Towards the "Invention" of the Melanesia/Polynesia Distinction 1595–1832', *Journal of Pacific History* (2003), vol. 38, no. 2, pp. 175–96, doi.org/10.1080/0022334032000120521).
7 Anaïs Maurer, *The Ocean on Fire: Pacific Stories from Nuclear Survivors and Climate Activists* (Duke University Press, 2024), p. 28.
8 Jaimey Hamilton Faris, 'Sisters of Ocean and Ice: On the Hydro-feminism of Kathy Jetñil-Kijiner and Aka Niviâna's Rise: From One Island to Another', *Shima* (2019), vol. 13, no. 2, p. 79, doi.org/10.21463/shima.13.2.08.
9 Maurer, *The Ocean on Fire*, p. 12.
10 Teresia K. Teaiwa, 'Bikinis and Other S/pacific N/oceans', *The Contemporary Pacific* (1994), vol. 6, no. 1, p. 87.
11 N. A. J. Taylor, 'The Antipodean Stance of Pam Debenham's 1980s Screenprints', in Livia Monnet (ed.), *Toxic Immanence: Decolonizing Nuclear Legacies and Futures* (McGill-Queen's University Press, 2022), pp. 316–18.
12 Tilman A. Ruff, 'The Humanitarian Impact and Implications of Nuclear Test Explosions in the Pacific Region', *International Review of the Red Cross* (2015), vol. 97, no. 899, p. 779, doi.org/10/1017/S1816383116000163.
13 Maurer, *The Ocean on Fire*, p. 11 and Teaiwa, 'Bikinis', p. 91.
14 Jack Niedenthal, *For the Good of Mankind: A History of the People of Bikini and Their Islands* (Bravo Publishers, 2013); Teaiwa, 'Bikinis', p. 91; Michelle Keown, '"A Story of a People on Fire": Nuclear Archives and Marshallese Cultural Memory in Kathy Jetñil-Kijiner's "Anointed"', in *Art and Australia Online*, 2020, https://www.artandaustralia.com/online/online/image-not-nothing-concrete-archives/%E2%80%98-story-people-fire%E2%80%99-nuclear-archives-and-marshallese.html [accessed 10 September 2024].
15 Nic Maclellan, 'The Nuclear Age in the Pacific Islands', *The Contemporary Pacific* (2005), vol. 17, no. 2, p. 363.

16 https://janecmi.com/See-Reverse-Side; Fiona Amundsen and Sylvia C. Frain, 'The Politics of Invisibility: Visualizing Legacies of Nuclear Imperialisms', *Journal of Transnational American Studies* (2020), vol. 11, no. 2, pp.131–5, doi.org/10.5070/T8112049588.
17 Kathy Jetñil Kijiner, 'Dome Poem Part III: "Anointed" Final Poem and Video', https://www.kathyjetnilkijiner.com/dome-poem-iii-anointed-final-poem-and-video/ [accessed 10 September 2024]; Keown, '"A Story of a People on Fire"'.
18 For more information, see https://www.aucklandmuseum.com/visit/galleries/ground-floor/tamaki-herenga-waka/pacific-sisters [accessed 10 September 2024].
19 'UN's Mission to Keep Plastics Out of Oceans and Marine Life', *UN News,* 27 April 2017, https://news.un.org/en/story/2017/04/556132-feature-uns-mission-keep-plastics-out-oceans-and-marine-life [accessed 10 September 2024].
20 Karen Jacobs, 'Bottled Ocean 2120: George Nuku, the Ocean, Plastic and the Role of Artists in Discussing Climate Change', *World Art* (2022), vol. 12, no. 3, pp. 213–38, doi.org/10.1080/21500894.2022.2070659.
21 Géraldine Le Roux, 'Transforming Representations of Marine Pollution. For a New Understanding of the Artistic Qualities and Social Values of Ghost Nets', *Anthrovision* (2016), vol. 4, no. 1, p. 1, doi.org/10.4000/anthrovision.2221.
22 Chris Parsons, 'The Pacific Islands: The Front Line in the Battle Against Climate Change', *US National Science Foundation*, 2022, https://new.nsf.gov/science-matters/pacific-islands-front-line-battle-against-climate [accessed 10 September 2024].
23 Margaret Jolly, 'Engendering the Anthropocene in Oceania: Fatalism, Resilience, Resistance', *Cultural Studies Review* (2019), vol. 25, no. 2, p. 172, doi.org/10.5130/csr.v25i2.6888.
24 Katerina Teaiwa, 'Our Rising Sea of Islands', *Pacific Studies* (2018), vol. 41, nos 1/2, pp. 26–54.
25 Veronica Strang, 'The Monstrous Sea', *Lagoonscapes: The Venice Journal of Environmental Humanities* (2023), vol. 3, no. 2, p. 260.
26 https://www.youtube.com/watch?v=rit6rRrruG8; Jaimey Hamilton Faris, 'Gestures of Survivance: Angela Tiatia's Lick and Feminist Environmental Performance Art in Oceania', *Pacific Arts* (2021), vol. 20, no. 1, p. 9.
27 https://350.org/rise-from-one-island-to-another/#poem; Hamilton Faris, 'Sisters of Ocean and Ice', p. 76.
28 Jacobs, 'Bottled Ocean 2120', p. 224.
29 https://www.nzatvenice.com/; https://paradisecamp.ws/. Unless otherwise specified, information on *Paradise Camp* was drawn from Yuki Kihara, *Paradise Camp* (Thames & Hudson Australia, 2022) and https://yukikihara.ws/; https://paradisecamp.ws/.
30 Kihara refers to the introduction of the New Zealand Crime Ordinance Act in 1961, which targeted the Fa'afafine community because it enforced laws that banned males impersonating females in public and a ban on homosexuality. However, following missionisation during the German colonisation of Sāmoa, strict gender binaries had already been imposed, resulting in discrimination of Fa'afafine and Fa'atama communities.
31 The country was a colony of the German Empire from 1899 to 1915, then came under a joint British and New Zealand colonial administration until 1 January 1962, when Sāmoa became independent.
32 Albert Wendt, 'Tatauing the Post-Colonial Body', New Zealand Electronic Poetry Centre, 1996, p. 42, https://www.nzepc.auckland.ac.nz/authors/wendt/tatauing.asp [accessed 10 September 2024]. Originally published in *Span* 42–43 (April-October 1996), pp. 15–29
33 *Paradise Camp*, https://paradisecamp.ws/videos-2921/ [accessed 10 September 2024].
34 Liang-Kai Yu and Eliza Steinbock, 'Yuki Kihara's *Paradise Camp* as a Potential Fa'afafine Museum: Fabulous Cohabitation in a Shared World', *Journal of Material Culture* (2023), vol. 28, no. 4, pp. 576–603 (p. 598), doi.org/10.1177/13591835231210440.
35 Epeli Hau'ofa, 'Our Sea of Islands', p. 160.

The Evolution of *Paradise Camp*

1 Ross Brooks, 'Darwin's Closet: The Queer Sides of The Descent of Man (1871)', *Zoological Journal of the Linnean Society* (9 January 2021), vol. 191, no. 2, doi.org/10.1093/zoolinnean/zlaa175.

2 Albert Wendt, Reina Whaitiri and Robert Sullivan, 'Introduction' in *Whetu Moana: Contemporary Polynesian Poems in English* (Auckland University Press, 2013), p. 10.

Estuarine

1 'Estuary' in *National Geographic*, https://education.nationalgeographic.org/resource/estuary/, [accessed 10 August 2024].

2 Micropolyphony is a musical texture developed by Hungarian-Austrian composer György Ligeti, consisting of many lines at different tempos or rhythms that merge into a unified harmonic sound, creating a dense texture.

3 Pierre Nora, 'Between Memory and History: Les Lieux de Mémoire', *Representations* (Spring 1989), no. 26, pp. 7–24; Susan Stewart, *On Longing: Narratives of the Miniature, the Gigantic, the Souvenir, the Collection* (Duke University Press, 1992).

4 Dino Carlo S. Santos, 'Our Lady of Peace and Good Voyage / Nuestra Señora de la Paz y Buen Viaje in *CCP Encyclopedia of Philippine Art Digital Edition*, 18 November 2020, https://epa.culturalcenter.gov.ph/3/82/2248/ [accessed 5 August 2024].

5 See Grace Odal, *Inang Tubig: Ang Diwa ng Ba'i sa Kalinangang Bayan* (Adhika ng Pilipinas, 1999); Danim R. Majerano, 'Ang Larawan ng Mutya: Muling-Pagsilang, Pagbabanyuhay', *Popularidad MALAY* (December 2023), vol. 36, no. 1, pp. 33–44.

6 See research of Agustin Sotto, https://manunuripelikula.com/1991-natatanging-gawad-urian/ [accessed 10 August 2024].

7 Michael Pante, 'The Esteros and Manila's Postwar Remaking', in Simo Laakkonen et al. (eds), *The Resilient City in World War II: Urban Environmental Histories* (Palgrave Macmillan, 2019), p. 238.

8 Anthony Medrano, 'Creole Waters: Stories of Culture and Ecology in Southeast Asia's Estuaries', Venice Architecture Biennale, 2023, https://www.youtube.com/watch?v=KAr7_gMC8V0 [accessed 11 September 2024].

9 Ibid.

10 Michelle Cacho Nicolasora, 'Kundiman: A Musical and Socio-cultural Exploration on the Development of the Philippine Art Song', University of Memphis, PhD thesis, 2014, p.102, https://digitalcommons.memphis.edu/etd/932/ [accessed 11 September 2024].

11 Ibid.

***Sea Inside:* Art and Marine Interiority**

1 Rachel Carson, 'Undersea', *Atlantic Monthly* (September 1937), https://www.theatlantic.com/magazine/archive/1937/09/undersea/652922/ [accessed 17 September 2024].

2 See for instance Stefan Helmreich, *Alien Ocean: Anthropological Voyages in Microbial Seas* (University of California Press, 2009).

3 Donna Haraway quoted in Melody Jue, *Wild Blue Media: Thinking Through Seawater* (Duke University Press, 2020), p. 9.

4 See for instance, Nancy Tuana, 'Porous Viscosity: Witnessing Katrina', in Stacy Alaimo and Susan Hekman (eds), *Material Feminisms* (Indiana University Press, 2008), pp. 188–213; Tiffany Lethabo King, *The Black Shoals: Offshore Formations of Black and Native Studies* (Duke University Press, 2019); Cecilia Chen, Janine MacLeod and Astrida Neimanis (eds), *Thinking With Water* (McGill-Queen's Press, 2013).

5 Stefanie Hessler (ed.), *Tidalectics: Imagining a Worldview Through Art and Science* (MIT Press, 2018) and Elizabeth Deloughrey and Tatiana Flores, 'Submerged Bodies: The Tidalectics of Representability and the Sea in Caribbean Art', in Lisa Blackmore and Liliana Gómez (eds), *Liquid Ecologies in Latin American and Caribbean Art* (Routledge, 2020), pp. 163–86.

6 Paul Gilroy, *The Black Atlantic: Modernity and Double Consciousness* (Verso, 1993); Dilip Menon et al., *Ocean as Method: Thinking with the Maritime* (Routledge, 2022).

7 Bronwyn Bailey-Charteris, *The Hydrocene: Eco-Aesthetics in the Age of Water* (Routledge, 2024), p. 106.

8 Alain Corbin, *The Lure of the Sea: The Discovery of the Seaside in the Western World 1750–1840*, trans. Jocelyn Phelps (Polity Press, 1994), pp. 1–18.
9 Rachel Carson, *The Sea Around Us* [1951] (Oxford University Press, 1961), p. 38.
10 Robin Jarvis, 'Hydromania: The Social History and Literary Significance of Romantic Swimming', in Margaret Cohen and Killian Quigley (eds), *The Aesthetics of the Undersea* (Routledge, 2019), pp. 67–82.
11 Margaret Cohen and Killian Quigley, 'Introduction: Submarine Aesthetics', in Cohen and Quigley, *The Aesthetics of the Undersea*, p. 2.
12 Lucy Lippard, *Undermining* (New Press, 2014), p. 88. See also Jasmine Liu, 'What do Native Artists Think of Michael Heizer's New Land Art Work?', *Hyperallergic* (September 2022), https://hyperallergic.com/763203/what-do-native-artists-think-of-michael-heizers-new-land-art-work/ [accessed 12 September 2024].
13 Alaina Claire Feldman, 'After Nature, Whose Ocean?', in Alaina Claire Feldman (ed.), *The Ocean After Nature* (Independent Curators International, 2016), p. 14.
14 See Francesca Curtis, 'Land Art Submerged: Betty Beaumont's *Ocean Landmark* and Art History Beyond Proximity', *Art History* (2024), vol. 47, no. 2, pp. 349 and 354.
15 Rachael Squire, 'Depth: Discovering, "Mastering", Exploring the Deep', in Kimberly Peters et al. (eds), *The Routledge Handbook of Ocean Space* (Routledge, 2022), pp. 353–6.
16 Teresia Teaiwa quoted in Epeli Hau'ofa, 'The Ocean in Us', *The Contemporary Pacific* (1998), vol. 10, no. 2, pp. 392–410.
17 Astrida Neimanis, *Bodies of Water: Posthuman Feminist Phenomenology* (Bloomsbury, 2017), p. 2.
18 Evan Ifekoya, 'Oceanic Feeling, Feeling Oceanic: Notes on Resonant Waves and Oceanic Sage', in *Practicing the Planetary* (Arts Catalyst, 2021), pp. 31–8.
19 Melody Jue and Rafico Ruiz, 'Thinking with Saturation Beyond Water: Thresholds, Phase Change, and the Precipitate', in Melody Jue and Rafico Ruiz (eds), *Saturation: An Elemental Politics* (Duke University Press, 2021), p. 3.
20 See Jules Michelet, *La Mer* (Hachette & Cie, 1861). Rachel Carson suggests that 'as life itself began in the sea, so each of us begins his [*sic*] individual life in a miniature ocean within his mother's womb, and in the stages of his embryonic development repeats the steps by which his race evolved, from gill-breathing inhabitants of a water world to creatures able to live on land'. Carson, *The Sea Around Us*, p. 8.
21 Neimanis, *Bodies of Water*, p. 3.
22 See Sabrina Imbler, *My Life in Sea Creatures* (Chatto & Windus, 2022), pp. 25–48.
23 As does the relationship of father and baby. Emanuele Coccia, *Metamorphoses* (Polity Press, 2021), p. 14.
24 Kasia Molga, 'How to Make an Ocean', in Pandora Syperek and Sarah Wade (eds), *Oceans: Documents of Contemporary Art* (Whitechapel Gallery/MIT Press, 2023), pp. 41–3.
25 *Chioma Ebinama: The Eyes of the Beloved are Everywhere*, Morena di Luna (1 July–10 September 2023), press release, https://www.maureenpaley.com/exhibitions/chioma-ebinama-The-Eyes-of-the-Beloved-are-Everywhere [accessed 17 September 2024].
26 Ibid.
27 Sunaura Taylor, *Disabled Ecologies: Lessons from a Wounded Desert* (University of California Press, 2024).
28 Julia Jung et al., 'Doubling Down on Wicked Problems: Ocean ArtScience Collaborations for a Sustainable Future', *Frontiers in Marine Science* (2022), vol. 9, doi.org/10.3389/fmars.2022.873990.
29 Astrida Neimanis, et al., 'Four Problems, Four Directions for Environmental Humanities: Toward Critical Posthumanities for the Anthropocene', *Ethics and the Environment* (2015), vol. 20, no. 1, pp. 67–97. *Project MUSE*, doi.org/10.2979/ethicsenviro.20.1.67.
30 Harun Morrison, 'Three New Sea Songs', in Kirsten Cooke (ed.), *Aqueous Humours: Fluid Ground* (Matt's Gallery and the Poorhouse Reading Rooms, 2025).
31 Celeste Olalquiaga, *The Artificial Kingdom: On the Kitsch Experience* (University of Minnesota Press, 1998).
32 In fact, John Miller locates an animal liberationist narrative behind the Victorian poet Eliza Keary's telling of the selkie myth in her 1874 poem 'Little Seal-Skin', https://sharc.sites.sheffield.ac.uk/blog/the-selkie-and-the-fur-trade-eliza-kearys-little-seal-skin [accessed 24 September 2024].

33 Tarralik Duffy, 'According to Shuvinai', in *Shuvinai Ashoona: Mapping Worlds* (The Power Plant Contemporary Art Gallery/Hirmer Verlag GmbH, 2021), pp. 63–8.
34 Personal conversation with the artist, 9 September 2024.
35 Pandora Syperek and Sarah Wade, 'Introduction: Towards an Oceanic Art History', in *Oceans: Documents of Contemporary Art* (Whitechapel Gallery/MIT Press, 2023), p. 15.
36 Ron Broglio, *Surface Encounters: Thinking with Animals and Art* (University of Minnesota Press, 2009), p. 125.
37 Nicola McCartney, 'Animal Drag: The Critical and Conscious Performance of Animality', *Humanimalia* (Spring 2024), vol. 14, no. 2, p. 265.
38 Timothy Morton, *Being Ecological* (Pelican Books, 2018), p. 186.
39 Jessica Ullrich, 'Animal Artistic Agency: Contemporary Interspecies Art and Relational Aesthetics', in Suzanne Anker and Suzanne Flach (eds), *The Cultures of Entanglement: On Nonhuman Forms in Contemporary Art* (transcript Verlag, 2024), pp. 149–64. Relatedly, Koen Taselaar's *RADICAL FURNITURE FOR RADICAL TIMES* (2019), featured in *A World of Water*, domesticates octopuses through representation in a tapestry showing these cephalopods lounging around on an array of furniture.
40 Marion Endt-Jones, 'A Monstrous Transformation: Coral in Art and Culture', in Marion-Endt Jones (ed.), *Coral: Something Rich and Strange* (Liverpool University Press, 2013), p. 17.
41 Pandora Syperek and Sarah Wade, 'Oceanic Curating', *Environmental Humanities* (November 2024), vol. 16, no. 3, p. 684.
42 Stephen C. Quinn, *Windows on Nature: The Great Habitat Dioramas of the American Museum of Natural History* (Abrams, 2006).

'The Ghost Whales'

1 Captain Paul Watson, 'The Blood of The Whales is on Danish Hands', *Ecologist*, 28 July 2015, https://theecologist.org/2015/jul/28/blood-whales-danish-hands [accessed 10 September 2024].

INDEX

NOTE: Page numbers in italic indicate an illustration.